THE DIVINE MILIEU

BY THE SAME AUTHOR

The Phenomenon of Man

Letters from a Traveller

Hymn of the Universe

The Future of Man

The Making of a Mind:
Letters from a Soldier-Priest: 1914–1919

The Appearance of Man

pierre teilhard de chardin

THE
DIVINE
MILIEU

HARPER PERENNIAL

NEW YORK • LONDON • TORONTO • SYDNEY • NEW DELHI • AUCKLAND

HARPER ● PERENNIAL

This book was originally published in French as *Le Milieu Divin* by Editions du Seuil, Paris, in 1957. This translation was originally published in Great Britain under the title *Le Milieu Divin*.

HarperCollins books may be purchased for educational, business, or sales promotional use. For information, please e-mail the Special Markets Department at SPsales@harpercollins.com.

First Harper Torchbook edition published 1965.
Revised Harper Torchbook edition published 1968.
First Perennial Classics edition published 2001.
Perennial Classics are published by Perennial, an imprint of HarperCollins Publishers.

Library of Congress Cataloging-in-Publication Data is available.
ISBN 0-06-093725-4

HB 03.06.2024

Contents

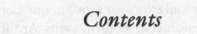

Note

⊚

Le Milieu Divin *is volume four in Pierre Teilhard de Chardin's* collected works as published in France, coming between *La Vision du Passé* and *L'Avenir de l'Homme*. In England and America it is volume two in the series, having been preceded by *The Phenomenon of Man* in 1959. If *The Phenomenon of Man* contained the kernel of Teilhard's scientific thought, *The Divine Milieu* is a key to the religious meditation that accompanied it.

All Teilhard's works involve grave problems for the translators, and the present version of *The Divine Milieu* is the result of much discussion and collaboration. Perhaps what most needs explanation is the retention of the word "milieu" from the original French title. This has been done more by necessity than by choice. The word "milieu" has no exact equivalent in English as it implies both centre and environment or setting: and even the normal use in England of the word "milieu" has insular associations. One suggested title, "In the Context of God," did not meet with the approval of the French committee in charge of the publication of Teilhard's works and I myself did not feel that another, "The Divine Environment," was close enough to the original. As we could reach no agreed solution, we left the word "milieu" in the title.

As a result of this, it was decided to retain the word "milieu" throughout the text also. Readers are asked to understand this word in the precise French connotation in which it was used by the author.

BERNARD WALL
General Editor of the English edition
of the works of Teilhard de Chardin

August 1960

SIC DEUS DILEXIT MUNDUM
For those who love the world

Foreword

@

Teilhard de Chardin: The Man
by Pierre Leroy, S.J.

The look in his eyes when they met your eyes revealed the man's
soul: his reassuring sympathy restored your confidence in
yourself. Just to speak to him made you feel better; you knew
that he was listening to you and that he understood you. His
own faith was in the invincible power of love: men hurt one
another by not loving one another. And this was not naïveté
but the goodness of the man, for he was good beyond the
common measure. In him, this belief was no mere conventional sentiment grafted on a generous nature, but the fruit
of long meditation; it was a certainty that came only with
years of reflection. It was this deep-seated spiritual conviction
that led Père Pierre Teilhard de Chardin to the practice of
self-forgetfulness: self being forgotten in a sympathetic union
with all men and with every individual man.

The combination of priest and scientist is nothing new;
but in his case what was really astonishing was his closeness
to the earth and his deep feeling for the value of matter. People who were shocked by him never realised how deep lay
the roots of this simultaneous love of God and of the world.
"Throughout my whole life," he wrote, "during every
moment I have lived, the world has gradually been taking on
light and fire for me, until it has come to envelop me in one
mass of luminosity, glowing from within . . . The purple flush

of matter fading imperceptibly into the gold of spirit, to be lost finally in the incandescence of a personal universe. . . .

"This is what I have learnt from my contact with the earth—the diaphany of the divine at the heart of a glowing universe, the divine radiating from the depths of matter a-flame."

There was something paradoxical in a priest who seemed outwardly so little the ecclesiastic, who was at home in even the least religious intellectual circles, who took his place in the advance-guard of thought, and devoted his life to the study of the properties of man as animal. It seemed paradoxical, too, that a specialist in the scientific history of the past should be interested only in the future.

He was all this: but above all he was a priest, deeply attached to the Church and its teaching, faithful to the end in spite of the annoyances and difficulties, the insinuations, too, that assailed him from every side.

Père Pierre Teilhard de Chardin lived during a period of doubt and perplexity. He witnessed the modernist crisis, with the sacrifices it entailed; he was driven from his own country by the injustice of political strife; and when he reached manhood he was caught up in the terrible war of 1914. A few years later he saw the collapse in the heat of revolution of social structures to which centuries of history seemed to have given permanence. He was present when forces were let loose which were to lead to a second world war; he was in Pekin when the atom bombs were dropped on Hiroshima and Nagasaki. It was his own fate to be misunderstood and condemned to silence, and to suffer torments that at times came near to overwhelming him. Like many others, he might well have retreated into his own solitary existence and abandoned his chosen field of activity, but his reaction was the exact opposite. In all that he did, as in all that he taught, there was no bitterness nor disillusioned cynicism, nothing but a constant optimism. Far from railing against the pettiness of men or the chaos of the world, he made it a rule never to assume the presence of evil. And when he was unable to deny the evidence of his eyes, he looked not for the damning but for the saving element in what he saw: a mental attitude that surely, if unexpectedly, provides the only road to truth.

This optimism had much more than a temperamental basis (of which we shall have more to say later); it was a conviction rooted deep in thought.

His scientific studies had taught Père Teilhard that the universe has its own history: it has a past, and it must be directed towards some final goal. "From the smallest individual detail to the vastest aggregations, our living universe (in common with our inorganic universe) has a structure, and this structure can owe its nature only to a phenomenon of growth." The world with all its riches, life with its astound ing achievements, man with the constant prodigy of his inventive powers, all are organically integrated in one single growth and one historical process, and all share the same upward progress towards an era of fulfilment. The inescapable dimension of time is a real function of growth and maturation, essential to our individual and collective becoming.

This growth must have some definite objective; there must be some term to the process: "The main stem of the tree of life," writes Père Teilhard, "has always climbed in the direction of the largest brain," towards, that is, greater spontaneity and greater consciousness.

Thus the slow progress of energies must reach a peak "from which life will never slip back." To overcome every obstacle, to unite our beings without loss of individual personality, there is a single force which nothing can replace and nothing destroy, a force which urges us forwards and draws us upwards: this is the force of love.

We can thus appreciate the central position in Père Teilhard's whole philosophy, of Christ, prototype of Man-Love; "God-Love reaching self-fulfilment only in love. Christianity," he tells us, "is nothing more nor less than a 'phylum of love' within nature."

Such, in his unshakable optimism and his passionate following of Christ, was Père Pierre Teilhard de Chardin. Today our minds are increasingly and agonisingly dismayed by the richness of matter and our own inability to find some coherent rule of conduct; our souls are perplexed by what we see happening around us and by the threat of what tomorrow may bring; and when, puzzled and terrified, we find a Christian with such confidence in the future both of man and of the world, we may tend to give a shrug of indifference and withdraw further into our shell of scepticism. Some of us will feel that the real, unhappily, can have nothing in common with the ideal; and it is to these disillusioned minds that the life of Père Teilhard provides an answer.

Much has already been written about him; his work as a scientist and his speculations in philosophy are beginning to be better known, but little has been said about what sort of a person he was in the daily business of life. It is not my intention, then, to analyse any particular aspect of his work, to deal with his scientific research or the expression he gave to his thought, nor to say any more about the difficulties he had to face and which in the end drove him into exile. My aim is rather to give a picture of the man I knew, to follow him through the different stages of his career and at the same time to try to read the secret that enabled him, in spite of the complications of his life, to achieve perfect interior unity.

During the years in which it was my good fortune, under

unusual conditions, to live close to him and work with him, I was able to some extent to decipher the mystery of his personality. I only hope that I may now succeed in explaining, in all humility, certain aspects of it and so make it easier to understand what lay behind the shining intelligence whose influence we now see to be so far-reaching.

The part of France in which he was born, Sarcenat, is a rugged country in which family life stood for a great deal. The family lived in an eighteenth-century manor-house, and the windows of the principal rooms gave on to a vista of volcanoes and rounded hills. From the top of the terrace, framed in greenery, you can see the capital of Auvergne, the vast plain of Clermont, with the foothills of the Puy mountains in the distance. No sound disturbs the tranquillity of the scene but the murmur of running water flowing with graceful constancy in a stone fountain.

Pierre Teilhard was born on 1 May, 1881, in a room on the first floor that looks out on the mountains. There used to be a very delicate pastel drawing hanging just opposite the left-hand window, which shows him as a curly-headed little boy, with the candid forehead and the thoughtful eyes that were his most striking features.

His childhood was that of an amenable little boy. "I was an affectionate child," he said, "good, and even pious." The predominating influence was that of his mother—his "dear, sainted maman"—to whom he owed "all that was best in his soul." "To rouse the fire into a blaze, a spark had to fall upon me; and the spark by which my own universe—still only halfway to being individually personalised—was to succeed in centring itself on its own fullness, undoubtedly came through my mother to light up and fire my child's soul."

His father, Emmanuel Teilhard de Chardin, was a man

who held strongly to a solid body of tradition, and he demanded from his children (Pierre was the fourth of eleven) active co-operation in a disciplined family life. Pierre owed to his father "more things than I could count," he wrote, "certain well-defined ambitions, no doubt, but even more a certain basic balance on which everything else was built."

Every day the household gathered in the dining-room after the evening meal to say prayers together. Pierre was to show me later, when I accompanied him to Sarcenat, his favourite place at prayers: he used to kneel by the wall while his parents rested their elbows on the table in the middle of the room.

The countryside is rich in rocks and minerals, in insect life and wild flowers, and M. Teilhard took pleasure in teaching his children how to understand and appreciate natural history. During their walks they would all gather mineral, zoological and botanical specimens and it was this interesting collection of local history that first encouraged Pierre's vocation as a scientist.

There was, however, another dominating interest that was typical of his temperament. He looked always for durability in his possessions and was not greatly attracted by the frail colouring of butterflies or the evanescent beauty of flowers. He has left a description of his feelings for what he calls his "idols": a plough-spanner carefully hidden in a corner of the courtyard, the top of a little metal rod, or some shell splinters picked up on a neighbouring range. "You should have seen me as in profound secrecy and silence I withdrew into the contemplation of my 'God of Iron,' delighting in its possession, gloating over its existence. A God, note, of iron; and why iron? Because in all my childish experience there was nothing in the world harder, tougher, more durable than this wonderful substance. There was about it a feeling of full personality, sharply individualised [. . .] But I can never forget the pathetic depths of a child's despair, when I realised one day that iron can be scratched and can rust [. . .] I had to look elsewhere for substitutes that would console me. Sometimes in the blue flame (at once so material, and yet so pure and intangible) flickering over the logs in the hearth, but more often in a more translucent and more delightfully coloured stone: quartz or amethyst crystals, and most of all glittering fragments of chalcedony such as I could pick up in the neighbourhood."

So we meet in his early youth the two components from which his whole life, both interior and in his relations with others, was to be built—a feeling for matter and a feeling for the durable.

He was sent to the Jesuit school at Villefranche; he was a good pupil, often at the top of his class, except in religious instruction. Not, indeed, that he was not ready to accept such teaching—far from it—but his mind seems to have instinctively reacted against the way in which it was taught. At that time the subject was still wrapped up in conventional phraseology and its presentation to children was dry and stodgy. Consider, for example, what Henri Brémond quotes in this respect: "Nothing is sweeter than to bask in the warmth that comes with the caress of grace. Jesus, my own brother, how well I know that the sweetest hours of my life are those of my monthly retreat, when I have you for my divine teacher; you open the book of my soul and help me to read about things that enrapture my powers. How good it is to go through this examination in love!" Poor children, adds Brémond; an examination, and, at that, an examination in love—what an attraction!—must have been the last straw.

It is not difficult to see that such "things" would hardly

"enrapture the powers" of a child like Pierre Teilhard, so eager for the permanent, the solid and the durable.

In any case, he must have set a good example to his companions, for we find him prefect (that is president) of the sodality, and deservedly looked up to in the school. His devotion to our Lady was tender and glowing, but it was now to gain in virility, and he was to assign to the Virgin Mary a dominant role in his concept of generative evolution.

At eighteen, after he had passed his baccalaureat, he said good-bye to his family and entered the Jesuit novitiate at Aix-en-Provence. Two years later he went to Laval to continue with his fellow-scholastics his studies in French, Latin and Greek. This was in 1902, when the religious orders were expelled from France; and he had to go abroad with the Community to seek refuge in Jersey.

He would often recall the comic epic of the move. In the hope of travelling unrecognised, the fathers wore civilian clothes. To give everyone a suit that fitted him would have been quite impossible, and so they had to make do with anything that came to hand. An appeal made to families brought the most varied collection of garments. Grave fathers and young scholastics found themselves donning a funeral top-hat with a light grey jacket, or a greenish old bowler with a long frock-coat, or a motoring-cap with a black morning coat; fifty years later Père Teilhard would still laugh at the memory of that masquerade.

The Jersey period was an important one. He studied scholastic philosophy, becoming familiar with its methods and terminology, without, however, adopting its spirit. He had an opportunity, at any rate, of turning over new problems in his mind, and was able to give some time to his favourite subject of geology. His contemporaries recall that

"Brother" Teilhard never went for a walk without his geologist's hammer and naturalist's magnifying glass.

In September 1905, after three years as a scholastic, he was sent to teach physics and chemistry at the Holy Family College in Cairo.

He was there for three years, during which he found time, in addition to his teaching, to deepen and extend his still imperfect knowledge of geology and palæontology. He was even able to publish in the scientific bulletin of Cairo a note on the Eocene in Upper Egypt, based on a collection he had made of its fossil fauna.

Egypt delighted his taste for the romantic. As he travelled along the Nile he must have dreamt, with the intense imagination revealed in his letters, of the exuberance of nature in unknown lands. "The East flowed over me in a first wave of exoticism: I gazed at it and drank it in eagerly—the country itself, not its peoples or its history (which as yet held no interest for me), but its light, its vegetation, its fauna and its deserts." Little did he anticipate that twenty-three years of his life were to be devoted to the Far East, to his "brooding old China."

At this point in his interior development, as we have seen, it was not man that attracted him. He had little interest in the peoples of the earth and their history. What drew him was nature in all its richness and diversity. The universe had taken bodily shape for him, but he had not yet become aware of its soul. Without realising it, Pierre Teilhard had reached a critical point in his life. He was more conscious than ever of the importance of the world, but it was a purely material world. He was in danger, if he was not careful, of succumbing to the lure of pantheism and losing himself in immensity: "to be all, one must be absorbed in all." "For only three years in Jersey," he writes, "and then for another three years in Cairo I studied (to the best of my ability) and taught (so far as my competence allowed me) fairly elementary physics, the pre-Quanta and pre-relativity physics of atomic structure: which means that in this subject I am an amateur, a mere layman. At the same time I find it difficult to express what a sense of fulfilment, ease, and of being at home I find in this world of electrons, nuclei and waves. If we wish to escape the inexorable fragility of the manifold, why not take refuge deeper, why not get beneath it? [. . .] Thus we may gain the world by renouncing it, by passively losing self in the heart of what has neither form nor dimension."

If we did not know that there may be contradictions in every person's make-up, we would fail to understand how Père Teilhard could have been tempted by the eastern "line," its self-centred passivity so foreign to his tastes. It is difficult to see how such an ardent and generous nature could have withdrawn from the contest. To let oneself be carried along passively in the cosmic eddy, to be lost in the intangible, seems completely inconsistent with a life already dedicated to action. It is important, therefore, to note that it was only at the speculative level that Père Teilhard contemplated the "eastern" solution: it had no influence on his faith as a Christian.

From Egypt he was sent to England for the last stages of his training as a priest and religious; and there a fuller and more satisfying view of the world forcibly impressed itself on him. It was then, one might say, that he began to direct his thought towards a philosophy of the person. The world now became for him a vast whole making its way towards a supreme personality; he had a vision of a universe in process of self-creation in which no breach could develop. He saw the image of the absolute reflected in the filigree of nature. "There were times when it really seemed to me that my eyes were about to see a universal being take shape in nature. Already, however, it was not by looking, as I used to look, for what is beyond matter that I sought to grasp and pin down the inexpressible ambience, but by looking for what is beyond the living."

The world, the whole universe, is an evolution—a genesis, to use Père Teilhard's own expression. Now every genesis presupposes inter-connections, mutual or reciprocal dependence, with no breach. It implies in the being that is forming itself a kinship between the composing elements; thus a static cosmos, fragmented in make-up, is unthinkable. If everything forms itself, everything must hold together. Matter and spirit, then, as we know them in our universe, are not two separate substances, set side by side and differing in nature. They are two distinct aspects of one single cosmic stuff and there is between them no conflict to baffle our intelligence. Physical energy contains in itself something of the spiritual, and since the upward trend of energy is a fact we can observe and verify with the increasing complexity of organisms, the law of the universe must surely be a continually progressing, irreversible, spiritualisation. Matter has now lost its former attraction for Père Teilhard: "the felicity that I had sought in iron, I can find now only in Spirit."

It was during these decisive years that he was ordained priest. Marked now with the priestly character, freed from the commitments involved in theological studies, and intellectually awake to the consequences of a generalised theory of evolution, he set about building up the structure of his own interior universe. This was now the pivot on which turned all his activities, his mental attitudes and his thought. He resolved in future to collaborate with all his energies in the cosmogenesis whose reality became for him daily more resplendent. Salvation was no longer to be sought in "abandoning the world" but in active "participation" in building it up. He would approach his scientific work no longer as an amateur but as a qualified specialist: and it would be undertaken not for its own sake (as he often insisted in conversation) but in order to release the Spirit from the crude ore in which it lay hidden or inactive.

The 1914 war did no more than delay his setting out on the great adventure of scientific research, which in his eyes was also the grand act of adoration. Obedient to the voice that called him, he was ready to plunge into another distressful adventure—an experience so monstrous and ghastly and yet, as we shall see, so exhilarating. He joined as a stretcher-bearer the 8th regiment of Moroccan Tirailleurs, later to become the 4th combined Tirailleurs and Zouaves. With the humble rank of corporal he was twice decorated, receiving the Médaille Militaire and the Légion d'Honneur.

He lived through the nightmare of war with all the generosity of his soul, with no thought for himself. Even amid scenes of death and devastation he was carried away by a sense of fulfilment. In war he breathed a new invigorating atmosphere. "The man at the front is no longer the same man." The shell of common assumptions and conventions was broken, and a fresh light shed on the hidden mechanism by which man's will has power to shape his development. Life takes on a new savour in the heroic devotion to a grand ideal. Père Teilhard felt that the reality he had found at the front would be with him for ever "in the great work of creation and of sanctifying humanity."

In 1919 he returned, this time for good, to his scientific career. He studied under Marcellin Boule at the Natural History Museum in Paris, and in 1922 his doctoral thesis was accepted. At the same time, somewhat against his own inclinations, he agreed to succeed Boussac, a son-in-law of P. Termier, who had been killed in the war, as professor of geology in the Institut Catholique in Paris. "Rather than this academic post," he wrote, "I should of course have preferred research work in Beyrouth or Shanghai or Trichinopoly, where men are needed." That those places should have come to his mind indicates the attraction to the tropics he had already begun to experience in Cairo. They were, in any case, familiar to him, for the French Jesuits had universities at both Beyrouth and Shanghai; and Trichinopoly was one of their important missions. It was natural, accordingly, that he should have thought of them as possible fields for a career in geology.

The brilliance of Père Teilhard's scientific and apostolic work was soon manifest. His influence was becoming increasingly felt in Paris both through his geological teaching and in his addresses to gatherings of Catholics at the École Normale, the Polytechnique, and the School of Mineralogy. The novelty and daring of his thought made a great appeal to the enthusiasm of young people eager to learn. It was at this moment, however, that an unexpected decision sent him to the Far East.

In 1914 one of his fellow-Jesuits, Père Émile Licent, had sailed for China with the intention of founding and building up a centre of scientific research into the natural resources of the Yellow River basin. For nine years he had been travelling over the great plain of Tchely, the Mongolian steppe which forms the Chinese border of the Tibetan plateau. He had been responsible for valuable collections bearing on its geology, botany, mineralogy and palæontology, and had built at Tientsin a museum and laboratory which he directed and inspired. In 1920 Père Licent had had one of those strokes of good fortune that sometimes attend workers in the field;

he had found a number of important fossiliferous deposits, and long camel trains had been carrying back through the provinces of Kansi and Shensi a precious store of mammiferous fossils collected in the Tertiary layers of the West.

In October 1921 Père Licent had sent the most significant of his acquisitions to the Paris Museum for expert appraisal. M. Boule entrusted the work to Père Teilhard, and a correspondence followed between the two Jesuits. In the end Père Teilhard agreed to join Père Licent in China and study the deposits on the spot. This was the origin of the "French Palæontological Mission," of which Père Licent was appointed director.

On 10 April, 1923, Père Teilhard embarked at Marseilles for Tientsin, where he arrived on the 23rd of the next month. He was now forty-two years old. He had volunteered for this distant assignment; his application had been favourably received, and everything seemed to augur a successful future. He was, in fact, soon to make one of the most important discoveries in his whole career. His personal impressions, however, on arriving, betray a surprising weariness. "I feel," he wrote on 27 May, "very much as though I had reached the limit of my powers: I seem somehow unable to keep things in my mind. I have a continual feeling that as far as my own life goes, the day is drawing to a close. The only way out, I think, is to cling to a blind and absolute faith in the meaning that all things—even the diminishments—must hold for a man who believes that God is the animating force behind every single event. The further I go, the more I am convinced that the only true science—the only one we can acquire in this ocean of weakness and ignorance—is the vision that begins to take shape under and through the multiplicity of things."

His weariness had something in common with that felt

by missionaries when they first come into contact with Northern China. Everything in the North is quite different from what they expected—flat, grey, dusty and nauseating. But Père Teilhard's confidence reveals another aspect of his nature: more than once the fineness of his feelings, his reserve towards others and even towards himself, checked the violence of his deeper emotions. Later he admitted the agonising distress that attacked and came close to overwhelming him; he lost confidence in himself; he was tortured by scruples; in spite of every effort of will he could not always disguise his suffering. Not unnaturally, therefore, there were times in his life when his friends noticed that he seemed to be abstracted and withdrawn.

Père Teilhard stayed for some weeks in Tientsin before setting out with Père Licent on an expedition to inner Mongolia and the Ordos desert. It was in this forgotten corner of the Chinese continent that they had the good fortune to find incontrovertible evidence for the existence of palæolithic man, hitherto unknown in these parts. Until that time nothing was known of pre-historic man south of the Yenisei. The discovery accordingly marked an essential step in the story of man.

Later in this volume the reader will find Père Teilhard's account of their adventures; but it may be interesting to note now his reaction to the East with which twenty years earlier he had wished to become acquainted. "I'm absorbed by the work, and very interested by the extreme novelty of what I'm seeing; interested, but not thrilled, as I would have been ten or twenty years ago. Today what counts for me (as for you) is the future of things [he is writing to Abbé Breuil] whereas here I am plunged into the past. Mongolia strikes me as a "museum" of antique specimens (zoological and ethnographical), a slice of the past. Try as I will, I see no promise of progress, no ferment, no 'burgeoning' for mankind of tomorrow. This corner of Asia (and even China outside the Great Wall) gives the impression of an empty reservoir."

We may set side by side of this a somewhat disillusioned passage from *Choses mongoles* which is conveniently included in this edition of the *Letters:* "It is a long time however, since I lost the illusion that travel brings us closer to the truth. . . . The more remote in time and space is the world we confront, the less it exists, and hence the poorer and more barren it is for our thought. So I have felt no disappointment this year at remaining quite untouched as I looked over the steppes where gazelles still run about as they did in the Tertiary period, or visited the yourts where the Mongols still live as they lived a thousand years ago. In what is, as in what was, there is nothing really new to be found."

It is easy to see how deeply Père Teilhard was imbued with his vision of working to build the future. Nothing else had the power to impress itself on him. And by the future he meant more than the building up of the material world; he envisaged the irreversible ascent, through men's efforts, to what he called the Omega Point.

It was during this expedition, in the stillness of the vast solitude of the Ordos desert, that one Easter Sunday he finished the mystical and philosophical poem, *Mass upon the altar of the World.* Alone before God, he prays with lyrical fervour: "Christ of glory, hidden power stirring in the heart of matter, glowing centre in which the unnumbered strands of the manifold are knit together; strength inexorable as the world and warm as life; you whose brow is of snow, whose eyes are of fire, whose feet are more dazzling than gold poured from the furnace; you whose hands hold captive the stars; you, the first and the last, the living, the dead, the re-born; you, who gather up in your superabundant oneness every delight, every taste, every energy, every phase of existence, to you my being cries out with a longing as vast as the universe: for you indeed are my Lord and my God."

When he got back to France in the autumn of 1924, an ordeal awaited him. Errors of theological interpretation had found their way into a note in which he expounded his new vision of the universe. His religious superiors had already taken alarm at the boldness of some of his philosophical views, which appealed particularly to the young, and thought it wise to bar him from teaching. Deeply wounded but submissive, he returned to China, where he became increasingly at home when he left the commercial and banking centre of Tientsin for the intellectual centre of Pekin.

It was a period of intense enthusiasm, and a succession of Chinese scientific institutions was coming into being, backed by eminent American and European scholars. Père Teilhard was completely at home in these exceptionally cosmopolitan circles. In particular he made friends with two outstanding characters: V. K. Ting, who had studied in Switzerland, and Wong Wen Hao, also with a European background (Belgium)—both geologists and both active in the new Chinese Geological Society. Ting was later mayor of Shanghai in Marshal Chiang Kai-Shek's administration; Wong also gave up science for politics and became Minister for Communications in Nationalist China.

On Sundays, there would be a gathering at the house of Dr. Grabau, the American palæontologist. Grabau was crippled with rheumatism and loved to see his friends around him. His lively intelligence, his kindness and the authority of his learning, carried his influence far and wide among the young intellectuals of Pekin. It was at his house that they discussed possible fields of study or publications that might be issued. I was greatly struck myself at these meetings by

Père Teilhard's winning manner. Buoyant and vivacious, with sufficient command of English to make jokes in it, he was, with our host, the life and soul of our gatherings. In addition to the originality of his thought and his personal charm, he had a quality rare in men of his stamp: he could listen to others and seem really interested in their suggestions. If they were too extravagant, he simply smiled.

The circle of friends whose common interests brought them to Pekin included men of international reputation in the world of science—Black, the Canadian, who was to be the first to publish an account of the Fossil-Man of Chou-Kou-Tien; Andersson, the prehistorian; Sven Hedin, the explorer; Granger and Barbour, the palæontologists; Chapman-Andrews, Höppeli and many others. First the war with Japan and then the civil war and setting-up of the Communist regime were to destroy the organisation they had so patiently and skilfully built up.

Meanwhile Père Teilhard had been renewing his contacts in France. It was in 1928, in his laboratory at the Paris Museum, that I met him for the first time. I had been chosen by my Jesuit superiors to work with Père Licent at the Tientsin Museum, and I was at the time reading for my degree at Nancy. When Père Teilhard came back to France in that year I met him at the Museum in the Place Valhubert. His simple and natural greeting immediately put me at my ease. He offered me a chair, while he sat casually on the edge of the table. His eyes, filled with intelligence and kindly understanding, his features, finely drawn and weathered by the winds of sea and desert, the glamour that surrounded his name, all made a deep impression on me. I can still hear the friendliness of his voice as he talked to me about China and the promising future, as it then appeared, that awaited it. For over an hour I listened to a flow of stimulating new ideas.

From that moment we were friends; and so we were to remain until the end.

His stay in France was brief. He went back that same year (1928) to Pekin, including on his way a visit to Ethiopia under the guidance of Henry de Monfreid. Later followed two important expeditions into Mongolia and Western China, and a return to Paris to help in organising the "Yellow Expedition."

It was in China that I met him again, in the spring of 1931. I had been working in the Gulf of Liao-Tong and on the Shantung peninsula. I had hopes of being back in Tientsin for Holy Week but shipping was constantly being delayed by the necessity to guard against the sudden attacks of the pirates who infested the area. My own journey, as it happened, was uneventful but it was late on Easter Sunday morning when I arrived at our house in Tientsin.

To my great joy, as I came in I met Père Teilhard in the corridor. He had been in Tientsin for some days and was on the point of leaving for Pekin, where he was to join the Citroën Central Asia Expedition (then held up at Kalgan by a serious breakdown). Teilhard, from his long experience of China, realised that my journey, made alone except for two Chinese servants, must have been one of great hardship; and my readjustment to normal life was made easier and quicker by his kind and tactful solicitude, exquisite tact and kindness. The next day we left together for Pekin. There I looked on while he identified the worked stones found in the jumble of fossils, and the deposits collected at Chou-Kou-Tien and placed in the Pekin Cenozoic Museum. These acquisitions held evidence that Sinanthropus might well have been responsible for their deliberate manufacture. The Abbé Breuil's more stringent examination was later to confirm the authenticity of the worked stones. Père Teilhard's keen powers of observation had not been mistaken. The rest of the story and the establishment of the near-certainty of Sinanthropus "Faber" is well known.

The Yellow Expedition was something of a disappointment. Père Teilhard's letters reveal the impatience of a geologist condemned to fill in time in order to forget the semicaptivity in which the whole party was kept by the hostility of the Chinese authorities.

Père Teilhard travelled later in India and several times visited America. During his brief returns to China between 1934 and 1938, he witnessed the disbanding of the Chinese national institutes he had seen come into being ten years earlier. His last work in the field was when he agreed to accompany some American friends in an investigation into the geology and prehistory of Burma. At the end of September 1938 he was back in Pekin; thence to Japan in the hope of salvaging a future that was daily more menaced. From Japan he sailed for Paris; and in August 1939, a few weeks before the declaration of war on Germany, he returned to Pekin.

War fastened its grip on Europe, but in China it stagnated for two years. The Japanese had occupied the North with brutal cynicism. Things were even more difficult in Tientsin, which suffered all the misery of arbitrary provocations. The territorial concessions (the districts reserved for Europeans and Americans) were isolated from the rest of the town by barbed-wire barricades guarded by arrogant sentries. The situation of the Tientsin Museum became precarious and it was decided to move it to Pekin. Père Teilhard agreed to spend some weeks in Tientsin helping to organise the moving of the collections built up during the last twenty-five years by Père Licent, whose health had obliged him to return to France for good. All we could do was to contrive to hang on in spite of everything and find some way to meet the difficulties that were aggravated for us as Frenchmen by the political and military situation.

The new house at Pekin, an annexe to the French barracks, was admirably organised. This became the Institute of Geobiology; and although it had not been designed for the purpose, it was pleasant, and life went on smoothly and comfortably.

Père Teilhard was in his office every morning at about eight o'clock, and we used to chat together for half an hour or so, seldom longer. After I had left him he used to jot down in an exercise book (over twenty of these survive) any comment of a philosophical, scientific or religious nature he thought worth preserving. These notes, written in diary form and with no apparent connecting thread, will enable students of Père Teilhard's thought to follow day by day, for over twenty years, the workings of a ceaselessly active mind. The time from nine o'clock to half-past twelve was devoted to writing his scientific papers and memoranda, occasionally to laboratory work; he preferred to spend his morning in writing and thinking. As far as possible we used to leave together in the early afternoon for the School of Medicine (the Pekin Union Medical College) to study its palæontological collection. Later, the whole of our time was spent at home, for all foreign institutions had been closed by the Japanese authorities. About five o'clock came visits to our friends. It was in these gatherings that you saw the real Père Teilhard; his mere presence brought an assurance of optimism and confidence. He had, too, the sort of mind that needs to retain and even multiply its contacts with the world outside; if he was to give substance to his thought or precision to his own personal ideas, he had to discuss his way of seeing things with other people.

Not that his conversation was always serious or pitched

on a high level. He was often, on the contrary, lively and gay; he appreciated good cooking and a good story; and sometimes his simplicity, or rather his unaffected frankness, could be embarrassing. Once, forgetting no doubt whom he was talking to, he embarked on an explanation that might have placed him in an awkward position. I was sitting beside him, and to attract his attention I nudged him gently with the tip of my toe. You can imagine my embarrassment when I heard him exclaim with a laugh: "Whose was that tactful kick?"

He had a fine sense of humour: his face would light up like a child's at a good joke; and if sometimes he could not resist an inviting target for his sly wit—after all, on his mother's side the blood of Voltaire flowed in his veins—it was done with such unaffected good humour that no-one could take it in bad part. It was one of his outstanding characteristics that he never gave way to bitterness, not even when decisions were taken that prevented the dissemination of his ideas. No wonder that he was universally loved and admired.

Père Teilhard was not, of course, without his opponents. It was not everyone that shared his optimism and broad-mindedness. There were some even who were irritated by it; for example, there was quite a little scene once when some remarks of Père Teilhard's caused an Ambassador to leave the table in the middle of a meal. As it happened, no harm was done, for Père Teilhard's simplicity and modesty made it easy to patch up the difference.

Living with Père Teilhard softened the harshness of our isolation, but one was sometimes conscious of how burdensome he found it to be confined within the walls of Pekin. It did violence to his nature to be thus sealed in, and his seeming gaiety was the fruit more of a victory of will and moral strength than of an inherent disposition.

Many have rightly been struck by Père Teilhard's great

optimism. He was indeed an optimist, in his attribution to the universe of a sense of direction in spite of the existence of evil and in spite of appearances; but in the daily life that concerned him personally, he was far from being an optimist. He bore with patience, it is true, trials that might well have proved too much for the strongest of us, but how often in intimate conversation have I found him depressed and with almost no heart to carry on. The agonising distress he already had to face in 1939 was intensified in the following years, and he sometimes felt that he could venture no further. During that period he was at times prostrated by fits of weeping, and he appeared to be on the verge of despair. But, calling on all the resources of his will, he abandoned himself to the supremely Great, to his Christ, as the only purpose of his being; and so hid his suffering and took up his work again, if not with joy, at least in the hope that his own personal vocation might be fulfilled.

Six years thus went by in the dispiriting atmosphere of China occupied by the Japanese and cut off from the rest of the world. In March 1946 Père Teilhard flew from Shanghai and there embarked in a ship that brought him back to France at the beginning of May. There he returned to his old room at Études, 15 rue Monsieur. His friends were quick to gather round him. Then, however, came a severe heart attack which struck him down just as he was on the point of leaving for a tour of South Africa: the similarity of the terrain in which the Australopithecidae and the Sinanthropus had been found had prompted the investigators in South Africa to ask for the assistance of an expert. It was two years, however, before Père Teilhard was able to make the journey. In 1951, after his election to the Académie des Sciences, he went to live in New York as a member of the Wenner Gren Foundation. There he devoted himself to anthropological studies, and became one of their most eminent associates.

Only once did he return to Paris, and then for a short visit. Although he always gave more than he received, he derived a new spiritual enrichment from this contact. It was during this last visit, in June 1954, that he expressed a desire to see the Lascaux caves. I had the pleasure of escorting him, and, as he had also to go to Lyons, our journey took us through Auvergne and we passed by his home at Sarcenat. Père Teilhard made no comment, but his silent absorption was sufficient indication of the memories evoked by these childhood scenes.

This was to be his last pilgrimage to France.

New restrictions had been imposed on him by his religious superiors, and it was broken by emotion he could hardly contain, and torn by unendurable anguish, that he cut short his stay and returned to New York six weeks earlier than he had intended.

In spite of the burden of spiritual distress, he took heart again and went back to work in his little New York office. I saw him for the last time a few days before Christmas. He was somewhat more tranquil, and was busy organising a scientific conference on anthropogenesis.

On 10 April, 1955, Easter Sunday, Père Teilhard collapsed to a sudden stroke just as he was about to have tea. He was walking over to the table when he fell like a stricken tree. For some moments there was an agonising silence and then he opened his eyes and said "What's happened—where am I?" When he was reassured he quietly uttered his last words, "This time, I feel it's terrible." He did not speak again. His doctor and his friend, Père de Breuvery, were sent for, but both were out. It was Fr. Martin Geraghty of St. Ignatius's, New York, who came immediately and administered Extreme Unction. The time was six o'clock in the evening. The sky was dazzling and spring was in its full splendour.

I was in Chicago at the time. I heard the news by telephone and hurried to New York, staying in the same hotel room he had slept in the night before. The whole staff was grief-stricken, from the humblest servant to the manager.

Père Teilhard's body lay in the chapel of the Jesuit house on Park Avenue, robed in his priest's vestments as he now lies for ever. He was hardly recognisable, the features drawn, the nose in sharp relief, the forehead smooth and unwrinkled. He reminded me of his compatriot from Clermont, Pascal.

The funeral was on Easter Tuesday, a grey, rainy day. Ten of his friends were present, but I was the only one to accompany him on the ninety-mile journey from New York to Saint Andrew on the Hudson. There he was buried, with a ceremony whose only distinction was its poverty, in the cemetery of the Jesuit novitiate for the New York Province.

It remains to consider a little more deeply the spiritual powers that provided the framework of Père Teilhard's complex existence and gave it cohesion; for what is important in a man is not so much what he achieves but the basic reason that inspires his activities. Since 1912, when he had completed his theological studies, Père Teilhard's aspirations were quite definitely formulated. He was a priest and a religious, and his first duty was therefore to Christ. At the same time he was resolved, as we have seen, to "participate" in the world, not to live in isolation from it. It was these seemingly contradictory principles, the service of Christ and participation in the world, that had to be reconciled. The evangelical doctrine of the Redemption through the Cross had to be reconciled with the salvation of the world through active co-operation in the building up of the universe.

This is not a problem of the speculative order, but one that calls for an immediate practical solution; and at its centre lies the question of the interior unity of life.

In examining Père Teilhard's answer, one point must be borne constantly in mind: he both accepts and practises the christian doctrine of detachment. He realises that the consummation of the world can be achieved only through a mystical death, a dark night, a renunciation of the whole being. So much we can take as established. But when he begins to look further into what constitutes renunciation, and to determine its mechanism, it may be held that he dissociates himself from ascetical practices hitherto accepted. His aim is to try out a new formula which, if it should prove effective, will enable men (already increasingly conscious of the tremendous impetus of technology) to look on Christianity not as a doctrine of impoverishment and diminution, but of expansion, and so to live as real Christians without ceasing to be artificers of the creative force. It matters little to him that God's omnipresent activity may appear as a differentiation or as a transformation, so long as Christ is attained and glorified: the problem is whether renunciation, conceived as a cutting-off of oneself from the world, is a practical proposition for the whole body of mankind.

In the life of each one of us, a vast area is occupied by the exertion of natural or social energies and it would be unfair to allow the value represented by these positive expressions of our activities to run to waste. It is not that Père Teilhard seeks to attach a permanent, absolute value to these various human achievements: he sees them as necessary stages through which the human group must pass in the course of its transformation. What interests him is not the particular form they take but the function they serve, and what matters is that not only the self-denial of the ascetic and the renunciation of the sufferer, but also our positive efforts to achieve natural perfection and to meet human obligations, should lead us to a consciousness of our spiritual growth.

Without, accordingly, sacrificing the mystical value of renunciation, it is seen to be essential to urge on the material development of the world with passionate conviction. Looked at from this angle, detachment and attachment can be harmonised and so complement one another. As he wrote to a friend who had the good fortune to see his business affairs prosper: "You are still having some difficulty in justifying to yourself the euphoria of a soul immersed in 'business.' I must point out to you that the really important thing is that you are actually experiencing that feeling of well-being. Bread was good for our bodies before we knew about the chemical laws of assimilation. . . . How, you ask, can the success of a commercial enterprise bring with it moral progress? And I answer, in this way, that since everything in the world follows the road to unification, the spiritual success of the universe is bound up with the correct functioning of every zone of that universe and particularly with the release of every possible energy in it. Because your enterprise (which I take to be legitimate) is going well, a little more health is being spread in the human mass, and in consequence a little more liberty to act, to think, and to love. . . . Because you are doing the best you can (though you may sometimes fail) you are forming your own self within the world, and you are helping the world to form itself around you."

Père Teilhard was fully alive to the danger that might lie in such statements. Wrongly interpreted or understood they might engulf the Christian in a type of pantheism that denied to the supernatural its pre-eminent position. He himself, firm in his faith in the universal value of creation, was in no danger of falling into this error. "I am not speaking metaphorically," he wrote, "when I say that it is throughout the length and breadth and depth of the world in movement that man can attain the experience and vision of his God." A critic by no means over-sympathetic to this line of thought, comments, "The driving force that runs through his thought and carries him along is that of a vigorous naturalism—impassioned and, without going so far as to say reckless, a little frightening."

It would indeed be frightening if one left out of account the underlying structure on which he built his search for God in and through his creatures: frightening, too, for a man ignorant of the laws of organic evolution and satisfied with the out-of-date concepts of a static world. There can be no doubt about the ambiguity of some of Père Teilhard's statements, for the very richness and originality of his thought made it difficult to express. He himself was always alive to this difficulty of expressing in adequate and unambiguous terms the vision of "a positive confluence of christian life with the natural sap of the universe." In his own self the integration of life had been achieved; if he loved God, it was through the world, and if he loved the world it was as a function of God, the animator of all things. "The joy and strength of my life," he wrote a month before his death, "will have lain in the realisation that when the two ingredients— God and the world—were brought together they set up an endless mutual reaction, producing a sudden blaze of such intense brilliance that all the depths of the world were lit up for me."

There was no contradiction in his soul, no ambiguity between his humble loyalty as a son of the Church and the boldness of his philosophical views. But in the depths of his being there raged the excruciating torment of reconciling his complete submission to the Church with the integrity of this thought.

In the following letter, written from Cape Town on 12 October, 1951, at the conclusion of his first visit to South Africa, Père Teilhard gives an excellent picture of his state of mind at that time and of the unreserved submission of his will to the decisions of the ecclesiastical authorities. It is addressed to his General, the Very Reverend Father Janssens, in Rome.

Cape Town, 12 October, 1951

Very Reverend Father,

P.C.

I feel that my departure from Africa (i.e., after two months' work and peace in the field) is a good moment to let you know briefly what I am thinking and where I stand. I do this without forgetting that you are the "General," but at the same time (as during our too short interview three years ago) with the frankness that is one of the Society's most precious assets.

1. Above all I feel that you must resign yourself to taking me as I am, that is, with the congenital quality (or weakness) which ever since my childhood has caused my spiritual life to be completely dominated by a sort of profound "feeling" for the organic realness of the World. At first it was an ill-defined feeling in my mind and heart, but as the years have gone by it has gradually become a precise, compelling sense of the Universe's general convergence upon itself; a convergence which coincides with, and culminates at its zenith in, him *in*

quo omnia constant, and whom the Society has taught me to love.

In the consciousness of this progression and synthesis of all things in *Xristo Jesu,* I have found an extraordinarily rich and inexhaustible source of clarity and interior strength, and an atmosphere outside which it is now physically impossible for me to breathe, to worship, to *believe.* What might have been taken in my attitude during the last thirty years for obstinacy or disrespect, is simply the result of my absolute inability to contain my own feeling of wonderment.

Everything stems from that basic psychological condition, and I can no more change it than I can change my age or the colour of my eyes.

2. Having made that clear, I can reassure you about my interior state of mind by emphasising that, whether or no this is generally true of others besides myself, the immediate effect of the interior attitude I have just described is to rivet me ever more firmly to three convictions which are the very marrow of Christianity.

The unique significance of Man as the spearhead of Life; the position of Catholicism as the central axis in the convergent bundle [*faisceau*] of human activities; and finally the essential function as consummator assumed by the risen Christ at the centre and peak of Creation: these three elements have driven (and continue to drive) roots so deep and so entangled in the whole fabric of my intellectual and religious perception that I could now tear them out only at the cost of destroying everything.

I can truly say—and this in virtue of the whole

structure of my thought—that I now feel more indissolubly bound to the hierarchical Church and to the Christ of the Gospel than ever before in my life. Never has Christ seemed to me more real, more personal or more immense.

How, then, can I believe that there is any evil in the road I am following?

3. I fully recognise, of course, that Rome may have its own reasons for judging that, in its present form, my concept of Christianity may be premature or incomplete and that at the present moment its wider diffusion may therefore be inopportune.

It is on this important point of formal loyalty and obedience that I am particularly anxious—it is in fact my real reason for writing this letter—to assure you that, in spite of any apparent evidence to the contrary, I am resolved to remain a "child of obedience."

Obviously I cannot abandon my own personal search—that would involve me in an interior catastrophe and in disloyalty to my most cherished vocation; but (and this has been true for some months) I have ceased to propagate my ideas and am confining myself to achieving a deeper personal insight into them. This attitude has been made easier for me by my now being once more in a position to do first-hand scientific work.

In fact I have every hope that my absence from Europe will allow the commotion about me that may have disturbed you recently, simply to die down. Providence seems to be lending me a helping hand towards this: what I mean is that the Wenner Gren (formerly the Viking) Foundation in New

York which sent me here (it is the same Foundation, incidentally, that refloated Père Schmidt's *Anthropos* after the war) is already asking me to prolong my stay in America as long as I can: they want me to classify and develop the data obtained from my work in Africa.

All this allows me a breathing space and gives a purely scientific orientation to the end of my career . . . and of my life.

Let me repeat that, as I see it, this letter is simply an exposition of conscience and calls for no answer from you. Look on it simply as a proof that you can count on me unreservedly to work for the kingdom of God, which is the one thing I keep before my eyes and the one goal to which science leads me.

Your most respectful *in Xto filius*

P. Teilhard de Chardin

Père Teilhard knew well that it was his duty to speak out and allow others to share the fruits of his own experience. "If I didn't write," he told me, "I would be a traitor." It was no doubt because he expressed himself with such frankness and unaffected simplicity that he met with so much opposition both from theologians and from scientists. Of the latter he wrote, "I have often felt myself impelled to question the value of my own interior testimony. Friends have assured me that they have never experienced this themselves. 'It's just a matter of temperament,' they've said. 'You feel the need to philosophise, while with us research is simply something we enjoy doing, like having a drink.' " Not a very convincing answer, it is true, and one that would by no means satisfy the scrupulous mind or soul of Père Teilhard, who felt it essential even in the conceptual order to justify his activity. "You fail," he replied, "to get to the bottom of what goes on in your heart and your mind, and that is why the 'cosmic sense' and faith in the world are still dormant in you. You may multiply the extent and duration of progress as much as you please, and you may promise the world another hundred million years of growth; but if at the end of that time it appears that the whole of consciousness must revert to zero without its hidden essence being anywhere preserved, then we shall lay down our arms and there will be a complete cessation of effort. The day is not far distant when humanity will realise that biologically it is faced with a choice between suicide and adoration."

Père Teilhard's life, his interior life, is thus seen as a witness. "My skill as a philosopher may be greater or less," he writes in his notes, "but one fact will always remain, that an average man of the twentieth century, just because he shared normally in the ideas and interests of his time, was able to attain a balanced interior life only in a scientifically integrated concept of the world and of Christ; and that therein he found peace and limitless scope for his being to expand. Today, my faith in God is sounder, and my faith in the world stronger, than ever." Could there be a more up-to-date or more faithful version of St. Paul's doctrine of the "cosmic" Christ? "In him all created things took their being, heavenly and earthly, visible, and invisible. . . . They were all created through him and in him; he takes precedence of all, and in him all subsist. . . . It was God's good pleasure to let all completeness dwell in him, and through him to win back all things, whether on earth or in heaven, unto union with himself, making peace with them through his blood, shed on the cross" (Coloss. I, 16–19, 20).

In the strength derived from the nobility of his task, he

could follow a road that might have led more ill-equipped souls into dangerous misconceptions; and this in all sincerity of conscience. It was no doubt because of this serenity that he was so tolerant, with a tolerance that bordered on weakness and often caused him to be misunderstood; for people are more ready to give others credit for justice than for love.

Even those who were most hostile to his philosophical and religious views recognised the exquisite gift for sympathy which made him a "catcher of souls." Countless intellectuals, executives, workmen and humble folk caught from him the vital spark of illumination and found peace. There was one limit to his tolerance: the one fault he detested, the one he would have nothing to do with, was the deliberate acceptance and delight in disgust with life, contempt for the works of man, fear of the human effort. For Père Teilhard this lack of confidence in the efficacy of man's vocation was the real sin. Our natural weaknesses could be looked on with indulgence, so long as the desire to "rise," to progress forward and upward, was sincere. "Anything that makes me sink lower—that," he used to say, "is the real evil."

It was this reasoned optimism, the fruit of his interior life, that gave him strength both when he had to fight and when he obediently gave way. It confirmed him, too, in his hope that one day the whole world would enrol in the service of Christ. From a continually reinvigorated search for God he drew fresh stores of tenacity. There was nothing petty nor rigid in this tension of the will towards union, but an effort renewed from day to day—God knows in the midst of what struggles—to steep himself in the divine presence, without which he counted everything as vanity; and at the same time he saw that, in pushing human aspirations to the most daring extremes, man may ascend to the heights. It was this he had in mind when he used to say, "We must dare all things."

He died suddenly, as he had prayed that he might, in the full vigour of life; friend of all men, of all countries, he died in the most cosmopolitan city in the world. It was on Easter Sunday, in the full bloom of spring, with the city bathed in a flood of sunshine. So it was that in the joy of the Resurrection Père Teilhard was reunited with the Christ whom all his life he had longed to possess in the blaze of victory.

> Lord, since with every instinct of my being and through all the changing fortunes of my life, it is you whom I have ever sought, you whom I have set at the heart of universal matter, it will be in a resplendence which shines through all things and in which all things are ablaze, that I shall have the felicity of closing my eyes.

THE DIVINE MILIEU

PREFACE

If the form and content of the following pages are to be rightly understood, the reader must not misconceive the spirit in which they were written.

This book is not specifically addressed to Christians who are firmly established in their faith and have nothing more to learn about its beliefs. It is written for the waverers, both inside and outside; that is to say for those who, instead of giving themselves wholly to the Church, either hesitate on its threshold or turn away in the hope of going beyond it.

As a result of changes which, over the last century, have modified our empirically based pictures of the world and hence the moral value of many of its elements, the "human religious ideal" inclines to stress certain tendencies and to express itself in terms which seem, at first sight, no longer to coincide with the "christian religious ideal."

Thus it is that those whose education or instinct leads them to listen primarily to the voices of the earth, have a certain fear that they may be false to themselves or diminish themselves if they simply follow the Gospel path.

So the purpose of this essay—on life or on inward vision—is to prove by a sort of tangible confirmation that this fear is unfounded, since the most traditional Christianity, expressed in Baptism, the Cross and the Eucharist, can be interpreted so as to embrace all that is best in the aspirations peculiar to our times.

My hope is that it may help to show that Christ, who is ever the same and ever new, has not ceased to be the "first" within mankind.

An Important Observation

The following pages do not pretend to offer a complete treatise on ascetical theology—they only offer a simple *description* of a *psychological* evolution observed *over a specified interval*. A possible series of inward perspectives gradually revealed to the mind in the course of a humble yet "illuminative" spiritual ascent—that is all we have tried to note down here.

The reader need not, therefore, be surprised at the apparently small space allotted to moral evil and sin: the soul with which we are dealing is assumed to have already turned away from the path of error.

Nor should the fact arouse concern that the action of grace is not referred to or invoked more explicitly. The subject under consideration is actual, concrete, "supernaturalised" man—but seen in the realm of *conscious* psychology only. So there was no need to distinguish explicitly between natural and supernatural, between divine influence and human operation. But although these technical terms are absent, the thing is everywhere taken for granted. Not only as a theoretically admitted entity, but rather as a living reality, the notion of grace impregnates the whole atmosphere of my book.

And in fact *the divine* milieu *would lose all its grandeur and all its savour* for the "mystic" if he did not feel—with his whole "participated" being, with his whole soul made receptive of the divine favour freely poured out upon it, with his whole will strengthened and encouraged—if he did not feel *so completely swept away* in the divine ocean that *no initial point of support* would be left him in the end, of his own, within himself, from which he could act.

INTRODUCTION

"In Eo Vivimus"

The enrichment and ferment of religious thought in our time has undoubtedly been caused by the revelation of the size and the unity of the world all around us and within us. All around us the physical sciences are endlessly extending the abysses of time and space, and ceaselessly discerning new relationships between the elements of the universe. Within us a whole world of affinities and interrelated sympathies, as old as the human soul, is being awakened by the stimulus of these great discoveries, and what has hitherto been dreamed rather than experienced is at last taking shape and consistency. Scholarly and discriminating among serious thinkers, simple or didactic among the half-educated, the aspirations towards a vaster and more organic *one,* and the premonitions of unknown forces and their application in new fields, are the same, and are emerging simultaneously on all sides. It is almost a commonplace today to find men who, quite naturally and unaffectedly, live in the explicit consciousness of being an atom or a citizen of the universe.

This collective awakening, similar to that which, at some given moment, makes each individual realise the true dimensions of his own life, must inevitably have a profound religious reaction on the mass of mankind—either to cast down or to exalt.

To some, the world had disclosed itself as too vast: within such immensity, man is lost and no longer counts; and there is nothing left for him to do but shut his eyes and disappear. To others, on the contrary, the world is too beautiful; and it, and it alone, must be adored.

There are Christians, as there are men, who remain unaffected by these feelings of anxiety or fascination. The following pages are not for them. But there are others who are alarmed by the agitation or the attraction invincibly produced in them by this new rising star. Is the Christ of the Gospels, imagined and loved within the dimensions of a Mediterranean world, capable of still embracing and still forming the centre of our prodigiously expanded universe? Is the world not in the process of becoming more vast, more close, more dazzling than Jehovah? Will it not burst our religion asunder? Eclipse our God?

Without daring, perhaps, to admit to this anxiety yet, there are many (as I know from having come across them all over the world) who nevertheless feel it deep within them. It is for those that I am writing.

I shall not attempt to embark on metaphysics or apologetics. Instead, I shall turn back, with those who care to follow me, to the Agora. There, in each other's company, we shall listen to St. Paul telling the Areopagites of "God, who made man that he might seek him—God whom we try to apprehend by the groping of our lives—that self-same God is as pervasive and perceptible as the atmosphere in which we are bathed. He encompasses us on all sides, like the world itself. What prevents you, then, from enfolding him in your arms? Only one thing: your inability *to see him*."*

This little book does no more than recapitulate the eternal lesson of the Church in the words of a man who, because he believes himself to feel deeply in tune with his own times, has sought to teach how to see God everywhere, to see him in all that is most hidden, most solid, and most ultimate in the world. These pages put forward no more than a practical attitude—or, more exactly perhaps, a way of teaching how to see. Let us begin by leaving argument aside for a moment. Place yourself here, where I am, and look from this privileged position—which is no hard-won height reserved for the elect, but the solid platform built by two thousand years of christian experience—and you will see how easily the two stars, whose divergent attractions were disorganising your faith, are brought into conjunction. Without mixture, without confusion, the true God, the christian God, will, under your gaze, invade the universe, our universe of today, the universe which so frightened you by its alarming size or its pagan beauty. He will penetrate it as a ray of light does a crystal; and, with the help of the great layers of creation, he will become for you universally perceptible and active—very near and very distant at one and the same time.

If you are able to focus your soul's eyes so as to perceive this magnificence, you will soon forget, I assure you, your

*At the end of his life the author reverted to *Le Milieu Divin* in two autobiographical works where he expands what he means by *seeing him*: "*Throughout* my life, *by means of* my life, the world has little by little caught fire in my sight until, aflame all around me, it has become almost completely luminous from within. . . . Such has been my experience in contact with the earth—the diaphany of the Divine at the heart of the universe on fire . . . Christ; his heart; a fire: capable of penetrating everywhere and, gradually, spreading everywhere." *French Editor's Note.*

unfounded fears in face of the mounting significance of the earth. Your one thought will be to exclaim: "Greater still, Lord, let your universe be greater still, so that I may hold you and be held by you by a contact at once made ever more intense and ever wider in its extent!"

The line we shall follow in our survey is quite simple. Since in the field of experience each man's existence can properly be divided into two parts—what he does and what he undergoes—we shall consider each of these parts in turn: the active and the passive. In each we shall find at the outset that, in accordance with his promise, God truly waits for us in things, unless indeed he advances to meet us. Next we shall marvel how the manifestation of his sublime presence in no way disturbs the harmony of our human attitude, but, on the contrary, brings it its true form and perfection. This done—that is, having shown that the two halves of our lives, and consequently the whole of our world, are full of God—it will remain for us to make an inventory of the wonderful properties of this *milieu* which is all around us (and which is nevertheless beyond and underlying everything), the only one in which, from now onwards, we are equipped to breathe freely.

PART ONE

The Divinisation of Our Activities

Note: It is of the utmost importance at this point to bear in mind what was said at the end of the Preface. We use the word "activity" in the ordinary, everyday sense, without in any way denying—far from it—all that occurs between *grace* and the *will* in the infra-experimental spheres of the soul. To repeat: what is most divine in God is that, in an absolute sense, we are nothing apart from him. The least admixture of what may be called Pelagianism would suffice to ruin immediately the beauties of the divine *milieu* in the eyes of the "seer."

Of the two halves or components into which our lives may be divided, the most important, judging by appearances and by the price we set upon it, is the sphere of activity, endeavour and development. There can, of course, be no action without reaction. And, of course, there is nothing in us which in origin and at its deepest is not, as St. Augustine said, *"in nobis, sine nobis."* When we act, as it seems, with the greatest spontaneity and vigour, we are to some extent led by the things we imagine we are controlling. Moreover, the very expansion of our energy (which reveals the core of our autonomous personality) is, ultimately, only our obedience to a will to be and to grow, of which we can master neither the varying intensity nor the countless modes. We shall return, at the beginning of Part Two, to these essentially passive elements, some of which form part of the very marrow of our being, while others are diffused among the inter-play of universal causes which we call our "character," our "nature" or our "good and bad luck." For the moment let us consider our life in terms of the categories and definitions which are the most immediate and universal. Everyone can distinguish quite clearly between the moments in which he is acting and those in which he is acted upon. Let us look at ourselves in one of those phases of dominant activity and try to see how, with the help of our activity and by developing it to the full, the divine presses in upon us and seeks to enter our lives.

1. THE UNDOUBTED EXISTENCE OF THE FACT AND THE DIFFICULTY OF EXPLAINING IT: THE CHRISTIAN PROBLEM OF THE SANCTIFICATION OF ACTION

Nothing is more certain, dogmatically, than that human action can be sanctified. "Whatever you do," says St. Paul, "do it in the name of our Lord Jesus Christ." And the dearest of christian traditions has always been to interpret those words to mean: in intimate union with our Lord Jesus Christ. St. Paul himself, after calling upon us to "put on Christ," goes on to forge the famous series of words *collaborare, compati, commori, con-ressuscitare,* giving them the fullest possible meaning, a literal meaning even, and expressing the conviction that every human life must—in some sort— become a life in common with the life of Christ. The actions of life, of which Paul is speaking here, should not, as everyone knows, be understood solely in the sense of religious and devotional "works" (prayers, fastings, alms-givings). It is the whole of human life, down to its most "natural" zones, which, the Church teaches, can be sanctified. "Whether you eat or whether you drink," St. Paul says. The whole history of the Church is there to attest it. Taken as a whole, then, from the most solemn declarations or examples of the pontiffs and doctors of the Church to the advice humbly given by the priest in confession, the general influence and practice of the Church has always been to dignify, ennoble and transfigure in God the duties inherent in one's station in life, the search for natural truth, and the development of human action.

The fact cannot be denied. But its legitimacy, that is its logical coherence with the whole basis of the christian temper, is not immediately evident. How is it that the perspectives opened up by the kingdom of God do not, by their very presence, shatter the distribution and balance of our activities? How can the man who believes in heaven and the Cross continue to believe seriously in the value of worldly occupations? How can the believer, in the name of everything that is most christian in him, carry out his duty as man to the fullest extent and as whole heartedly and freely as if he were on the direct road to God? That is what is not altogether clear at first sight; and in fact disturbs more minds than one thinks.

The question might be put in this way:

According to the most sacred articles of his *Credo,* the Christian believes that life here below is continued in a life of which the joys, the sufferings, the reality, are quite incommensurable with the present conditions in our universe. This contrast and disproportion are enough, by themselves, to rob us of our taste for the world and our interest in it; but to them must be added a positive doctrine of judgement upon, even disdain for, a fallen and vitiated world. "Perfection consists in detachment; the world around us is vanity and ashes." The believer is constantly reading or hearing these austere words. How can he reconcile them with that other counsel, usually coming from the same master and in any case written in his heart by nature, that he must be an example unto the Gentiles in devotion to duty, in energy, and even in leadership in all the spheres opened up by man's activity? There is no need for us to consider the wayward or the lazy who cannot be bothered to acquire an understanding of their world, or seek with care to advance their fellows' welfare—from which they will benefit a hundredfold after their last breath—and only contribute to the human task "with the tips of their fingers." But there is a kind of human spirit (known to every spiritual director) for whom this difficulty assumes the shape and importance of a besetting and numbing uncertainty. Such spirits, set upon interior unity, become the victims of a veritable spiritual dualism. On the one hand a very sure instinct, mingled with their love for that which is, and their taste for life, draws them to the joy of creating and of knowing. On the other hand a higher will to love God above all else makes them afraid of the least division or deflection in their allegiances. In the most spiritual layers of their being they experience a tension between the opposing ebb and flow caused by the drawing power of the two rival stars we spoke of at the beginning: God and the world. Which of the two is to make itself more nobly adored?

Depending on the greater or less vitality of the nature of the individual, this conflict is in danger of finding its solution in one of the three following ways: either the Christian will repress his taste for the tangible and force himself to confine his concern to purely religious objects, and he will try to live in a world that he has divinised by banishing the largest possible number of earthly objects; or else, harassed by that inward conflict which hampers him, he will dismiss the evangelical counsels and decide to lead what seems to him a complete and human life; or else, again, and this is the most usual case, he will give up any attempt to make sense of his situation; he will never belong wholly to God, nor ever wholly to things; incomplete in his own eyes, and insincere in the eyes of his fellows, he will gradually acquiesce in a double life. I am speaking, it should not be forgotten, from experience.

For various reasons, all three of these solutions are to be feared. Whether we become distorted, disgusted, or divided, the result is equally bad, and certainly contrary to that which Christianity should rightly produce in us. There is, without possible doubt, a fourth way out of the problem: it consists in seeing how, without making the smallest concession to "nature" but with a thirst for greater perfection, we can reconcile, and provide mutual nourishment for, the love of God and the healthy love of the world, a striving towards detachment and a striving towards the enrichment of our human lives. . . .

Let us look at the two solutions that can be brought to the christian problem of "the divinisation of human activity," the first partial, the second complete.

2. AN INCOMPLETE SOLUTION: HUMAN ACTION HAS NO VALUE OTHER THAN THE INTENTION WHICH DIRECTS IT

If we try somewhat crudely to reduce to its barest bones the immediate answer given by spiritual directors to those who ask them how a Christian, who is determined to disdain the world and jealously to keep his heart for God, can love what he is doing (his work)—in conformity with the Church's teaching that the faithful should take *not a lesser* but a *fuller* part than the pagan—it will run along these lines:

> You are anxious, my friend, to restore its value to your human endeavour; to you the characteristic viewpoints of christian asceticism seem to set far too little store by such activity. Very well then, you must let the clear spring water of purity of intention flow into your work, as if it were its very substance. Cleanse your intention, and the least of your actions will be filled with God. Certainly the material side of your actions has no definitive value. Whether men discover one truth or one fact more or less, whether or not they make beautiful music or beautiful pictures, whether their organisation of the world is more or less success-ful—all that has no direct importance for heaven. None of these discoveries or creations will become one of the stones of which is built the New Jerusalem. But what *will* count, up there, what *will* always endure, is this: that you have acted in all things *conformably* to the will of God.
>
> God obviously has no need of the products of your busy activity, since he could give himself every-thing without you. The only thing that concerns him, the only thing he desires intensely, is your faithful use of your freedom, and the preference you accord him over the things around you.
>
> Try to grasp this: the things which are given to you on earth are given you purely as an exercise, a "blank sheet" on which you make your own mind and heart. You are on a testing-ground where God can judge whether you are capable of being translated to

heaven and into his presence. You are on trial. So that it matters very little what becomes of the fruits of the earth, or what they are worth. The whole question is whether you have used them in order to learn how to obey and how to love.

You should not, therefore, set store by the coarse outer-covering of your human actions: this can be burnt like straw or smashed like china. Think, rather, that into each of these humble vessels you can pour, like a sap or a precious liquor, the spirit of obedience and of union with God. If worldly aims have no value in themselves, you can love them for the opportunity they give you of proving your faithfulness to God.

We are not suggesting that the foregoing words have ever been actually used; but we believe they convey a nuance which is often discernible in spiritual direction, and we are sure that they give a rough idea of what a good number of the "directed" have understood and retained of the exhortations given them.

On this assumption let us examine the attitude which they recommend.

In the first place this attitude contains an enormous part of truth. It is perfectly right to exalt the role of a good intention as the necessary start and foundation of all else; indeed—a point which we shall have to make again—it is the golden key which unlocks our inward personal world to God's presence. It expresses vigorously the primary worth of the divine will which, by virtue of this attitude, becomes for the Christian (as it was for his divine model) the fortifying marrow of all earthly nourishment. It reveals a sort of unique *milieu*, unchanging beneath the diversity and number of the tasks which, as men and women, we have to do, in which we can place ourselves without ever having to withdraw.

These various features convey a first and essential approximation to the solution we are looking for; and we shall certainly retain them in their entirety in the more satisfactory plan of the interior life which will soon be suggested. But they seem to us to lack the achievement which our spiritual peace and joy so imperiously demand. The divinisation of our endeavour by the value of the intention put into it, pours a priceless *soul* into all our actions; but *it does not confer the hope of resurrection upon their bodies.* Yet that hope is what we need if our joy is to be complete. It is certainly a very great thing to be able to think that, if we love God, something of our inner activity, of our *operatio,* will never be lost. But will not the work itself of our minds, of our hearts, and of our hands—that is to say, our achievements, what we bring into being, our *opus*—will not this, too, in some sense be "eternalised" and saved?

Indeed, Lord, it will be—by virtue of a claim which you yourself have implanted at the very centre of my will! I desire and need that it should be.

I desire it because I love irresistibly all that your continuous help enables me to bring each day to reality. A thought, a material improvement, a harmony, a unique nuance of human love, the enchanting complexity of a smile or a glance, all these new beauties that appear for the first time, in me or around me, on the human face of the earth—I cherish them like children and cannot believe that they will die entirely in their flesh. If I believed that these things were to perish for ever, should I have given them life? The more I examine myself, the more I discover this psychological truth: that no one lifts his little finger to do the smallest task unless moved, however obscurely, by the conviction that he is contributing infinitesimally (at least indirectly) to the building of something definitive—that is to say, to your work, my God. This may well sound strange or exagger-

ated to those who act without thoroughly scrutinising themselves. And yet it is a fundamental law of their action. It requires no less than the pull of what men call the Absolute, no less than you yourself, to set in motion the frail liberty which you have given us. And that being so, everything which diminishes my explicit faith in the heavenly value of the results of my endeavour, diminishes irremediably my power to act.

Show all your faithful, Lord, in what a full and true sense "their work follows them" into your kingdom—opera sequuntur illos. Otherwise they will become like those idle workmen who are not spurred by their task. And even if a sound human instinct prevails over their hesitancies or the sophisms of an incompletely enlightened religious practice, they will remain fundamentally divided and frustrated; and it will be said that the sons of heaven cannot compete on the human level, in conviction and hence on equal terms, with the children of the world.

3. THE FINAL SOLUTION: ALL ENDEAVOUR COOPERATES TO COMPLETE THE WORLD IN CHRISTO JESU

The general ordering of the salvation (which is to say the divinisation) of what we do can be expressed briefly in the following syllogism.

At the heart of our universe, each soul exists for God, in our Lord.

But all reality, even material reality, around each one of us, exists for our souls.

Hence, all sensible reality, around each one of us, exists, through our souls, for God in our Lord.

Let us examine each proposition of the syllogism in turn

and separately. Its terms and the link between them are easy to grasp. But we must beware: it is one thing to have understood its words, and another to have penetrated the astonishing world whose inexhaustible riches are revealed by its calm and formal exactitude.

A. *At the heart of our universe, each soul exists for God in our Lord*

The major of the syllogism does no more than express the fundamental Catholic dogma which all other dogmas merely explain or define. It therefore requires no proof here; but it does need to be strictly understood by the intelligence. Each soul exists for God in our Lord. We should not be content to give this destination of our being in Christ a meaning too obviously modelled on the legal relationships which in our world link an object to its owner. Its nature is altogether more physical and deeper. Because the consummation of the world (what Paul calls the Pleroma) is a communion of persons (the communion of saints), our minds require that we should express the links within that communion by analogies drawn from society. Moreover, in order to avoid the perverse pantheism and materialism which lie in wait for our thought whenever it applies to its mystical concepts the powerful but dangerous resources of analogies drawn from organic life, the majority of theologians (more cautious on this point than St. Paul) do not favour too realist an interpretation of the links which bind the limbs to the head in the Mystical Body. But there is no reason why caution should become timidity. If we want a full and vivid understanding of the teachings of the Church (which alone makes them beautiful and acceptable) on the value of human life and the promises or threats of the future life—then, without rejecting anything of the forces of freedom and of consciousness which form the natural endowment proper to the human soul, we must perceive the existence of links between us and the Incarnate Word no less precise than those which control, in the world, the affinities of the elements in the building up of "natural" wholes.

There is no point, here, in seeking a new name by which to designate the super-eminent nature of that dependence, where all that is most flexible in human combinations and all that is most intransigent in organic structures, merge harmoniously in a moment of final incandescence. We will continue to call it by the name that has always been used: *mystical* union. Far from implying some idea of diminution, we use the term to mean the strengthening and purification of the reality and urgency contained in the most powerful interconnections revealed to us in every order of the physical and human world. On that path we can advance without fear of over-stepping the truth; for everyone in the Church of God is agreed upon the fact itself, if not upon its systematic statement: by virtue of the powerful incarnation of the Word, our soul is wholly dedicated to Christ and centred upon him.

B. *"In our universe," we went on to say, "in which each soul exists for God, in our Lord, all that is sensible, in its turn, exists for the soul."*

In the form in which we have given it, the minor of our syllogism is tinged with a certain "finalist" doctrine which may shock those with a positivist cast of mind. Nevertheless it does no more than express an incontrovertible natural fact—which is that our spiritual being is continually nourished by the countless energies of the perceptible world. Here, again, proof is unnecessary. But it is essential to see—to see things as they are and to see them really and intensely. We live at the centre of the network of cosmic influences as we live at the heart of the human crowd or among the myriads of stars, without, alas, being aware of their immensity. If we wish to live our humanity and our Christianity to the full, we must overcome that insensitivity which tends to conceal things from us in proportion as they are too close to us or too vast. It is worth while performing the salutary exercise which consists in starting with those elements of our conscious life in which our awareness of ourselves as persons is most fully developed, and moving out from these to consider the spread of our being. We shall be astonished at the extent and the intimacy of our relationship with the universe.

Where are the roots of our being? In the first place they plunge back and down into the unfathomable past. How great is the mystery of the first cells which were one day animated by the breath of our souls! How impossible to decipher the welding of successive influences in which we are for ever incorporated! In each one of us, through matter, the whole history of the world is in part reflected. And however autonomous our soul, it is indebted to an inheritance worked upon from all sides—before ever it came into being—by the totality of the energies of the earth: it meets and rejoins life at a determined level. Then, hardly has it entered actively into the universe at that particular point than it feels, in its turn, besieged and penetrated by the flow of cosmic influences which have to be ordered and assimilated. Let us look around us: the waves come from all sides and from the farthest horizon. Through every cleft the world we perceive floods us with its riches—food for the body, nourishment for the eyes, harmony of sounds and fullness of the heart, unknown phenomena and new truths, all these treasures, all these stimuli, all these calls, coming to us from the four corners of the world, cross our consciousness at every moment.

What is their role within us? What will their effect be, even if we welcome them passively or indistinctly, like bad workmen? They will merge into the most intimate life of our soul and either develop it or poison it. We only have to look at ourselves for one moment to realise this, and either feel delight or anxiety. If even the most humble and most material of our foods is capable of deeply influencing our most spiritual faculties, what can be said of the infinitely more penetrating energies conveyed to us by the music of tones, of notes, of words, of ideas? We have not, in us, a body which takes its nourishment independently of our soul. Everything that the body has admitted and has begun to transform must be transfigured by the soul in its turn. The soul does this, no doubt, in its own way and with its own dignity. But it cannot escape from this universal contact nor from that unremitting labour. And that is how the characteristic power of understanding and loving, which will form its immaterial individuality, is gradually perfected in it for its own good and at its own risk. We hardly know in what proportions and under what guise our natural faculties will pass over into the final act of the vision of God. But it can hardly be doubted that, with God's help, it is here below that we give ourselves the eyes and the heart which a final transfiguration will make the organs of a power of adoration, and of a capacity for beatification, particular to each individual man and woman among us.

The masters of the spiritual life incessantly repeat that God wants only souls. To give those words their true value, we must not forget that the human soul, however independently created our philosophy represents it as being, is inseparable, in its birth and in its growth, from the universe into which it is born. In each soul, God loves and partly saves the whole world which that soul sums up in an incommunicable and particular way. But this summing-up, this welding, are not given to us ready-made and complete with the first awakening of consciousness. It is we who, through our own activity, must industriously assemble the widely scattered elements. The labour of seaweed as it concentrates in its tissues the substances scattered, in infinitesimal quantities, throughout the vast layers of the ocean; the industry of bees as they make honey from the juices broadcast in so many flowers—these are but pale images of the ceaseless working-over that all the forces of the universe undergo in us in order to reach the level of spirit.

Thus every man, in the course of his life, must not only show himself obedient and docile. By his fidelity he must *build*—starting with the most natural territory of his own self—a work, an *opus*, into which something enters from all the elements of the earth. *He makes his own soul* throughout all his earthly days; and at the same time he collaborates in another work, in another *opus*, which infinitely transcends, while at the same time it narrowly determines, the perspectives of his individual achievement: the completing of the world. For in presenting the christian doctrine of salvation, it must not be forgotten that the world, taken as a whole, that is to say in so far as it consists in a hierarchy of souls—which appear only successively, develop only collectively and will be completed only in union—the world, too, undergoes a sort of vast "ontogenesis" (a vast becoming what it is) in which the development of each soul, assisted by the perceptible realities on which it depends, is but a diminished harmonic. Beneath our efforts to put spiritual form into our own lives, the world slowly accumulates, starting with the whole of matter, that which will make of it the Heavenly Jerusalem or the New Earth.

C. *We can now bring together the major and minor*
of our syllogism so as to grasp the link between
them and the conclusion

If it is true, as we know from the Creed, that souls enter so
intimately into Christ and God, and if it is true, as we know
from the most general conclusions of psycho-analysis, that
the perceptible enters vitally into the most spiritual zones of
our souls—then we must also recognise that in the whole
process which from first to last activates and directs the elements of the universe, *everything forms a single whole*. And
we begin to see more distinctly the great sun of Christ the
King, of Christ *amictus mundo,* of the universal Christ, rising
over our interior world. Little by little, stage by stage, everything is finally linked to the supreme centre *in quo omnia*
constant. The streams which flow from this centre operate
not only within the higher reaches of the world, where
human activities take place in a distinctively supernatural and
meritorious form. In order to save and establish these sublime forces, the power of the Word Incarnate penetrates matter itself; it goes down into the deepest depths of the lower
forces. And the Incarnation will be complete only when the
part of chosen substance contained in every object—given
spiritual import once in our souls and a second time with
our souls in Jesus—shall have rejoined the final centre of its
completion. *Quid est quod ascendit, nisi quod prius descendit,*
ut repleret omnia?

It is through the collaboration which he stimulates in us
that Christ, starting from *all* created things, is consummated
and attains his plenitude. St. Paul himself tells us so. We may,
perhaps, imagine that the creation was finished long ago. But
that would be quite wrong. It continues still more magnifi-
cently, and at the highest levels of the world. *Omnis creatura*

adhuc ingemiscit et parturit. And we serve to complete it, even by the humblest work of our hands. That is, ultimately, the meaning and value of our acts. Owing to the interrelation between matter, soul and Christ, we bring part of the being which he desires back to God *in whatever we do.* With each one of our *works,* we labour—in individual separation, but no less really—to build the Pleroma; that is to say, we bring to Christ a little fulfilment.

4. COMMUNION THROUGH ACTION

Each one of our works, by its more or less remote or direct effect upon the spiritual world, helps to make perfect Christ in his mystical totality. That is the fullest possible answer to the question: How can we, following the call of St. Paul, see God in all the active half of our lives? In fact, through the unceasing operation of the Incarnation, the divine so thoroughly permeates all our creaturely energies that, in order to meet it and lay hold on it, we could not find a more fitting setting than that of our action.

To begin with, in action I adhere to the creative power of God; I coincide with it; I become not only its instrument but its living extension. And as there is nothing more personal in a being than his will, I merge myself, in a sense, through my heart, with the very heart of God. This commerce is continuous because I am always acting; and at the same time, since I can never set a boundary to the perfection of my fidelity nor to the fervour of my intention, this commerce enables me to liken myself, ever more strictly and indefinitely, to God.

The soul does not pause to relish this communion, nor does it lose sight of the material end of its action; for it is wedded to a *creative* effort. The will to succeed, a certain passionate delight in the work to be done, form an integral part of our creaturely fidelity. It follows that the very sincerity with which we desire and pursue success for God's sake reveals itself as a new factor—also without limits—in our being knit together with him who animates us. Originally we had fellowship with God in the simple common exercise of wills; but now we unite ourselves with him in the shared love of the end for which we are working; and the crowning marvel is that, with the possession of this end, we have the utter joy of discovering his presence once again.

All this follows directly from what was said a moment back on the relationship between natural and supernatural actions in the world. Any increase that I can bring upon myself or upon things is translated into some increase in my power to love and some progress in Christ's blessed hold upon the universe. Our work appears to us, in the main, as a way of earning our daily bread. But its essential virtue is on a higher level: through it we complete in ourselves the subject of the divine union; and through it again we somehow make to grow in stature the divine term of the one with whom we are united, our Lord Jesus Christ. Hence whatever our role as men may be, whether we are artists, working-men or scholars, we can, if we are Christians, speed towards the object of our work as though towards an opening on to the supreme fulfilment of our beings. Indeed, without exaggeration or excess in thought or expression—but simply by confronting the most fundamental truths of our faith and of experience—we are led to the following observation: God is inexhaustibly attainable in the *totality* of our action. And this prodigy of divinisation has nothing with which we dare to compare it except the subtle, gentle sweetness with which this actual change of shape is wrought; for it is achieved without disturbing at all *(non minuit, sed sacravit . . .)* the completeness and unity of man's endeavour.

5. THE CHRISTIAN PERFECTION OF HUMAN ENDEAVOUR

There was reason to fear, as we have said, that the introduction of christian perspectives might seriously upset the ordering of human action; that the seeking after, and waiting for, the kingdom of heaven might deflect human activity from its natural tasks, or at least entirely eclipse any interest in them. Now we see why this cannot and must not be so. The knitting together of God and the world has just taken place under our eyes in the domain of action. No, God does not deflect our gaze prematurely from the work he himself has given us, since he presents himself to us as attainable through that very work. Nor does he blot out, in his intense light, the detail of our earthly aims, since the closeness of our union with him is in fact determined by the exact fulfilment of the least of our tasks. We ought to accustom ourselves to this basic truth till we are steeped in it, until it becomes as familiar to us as the perception of shape or the reading of words. God, in all that is most living and incarnate in him, is not far away from us, altogether apart from the world we see, touch, hear, smell and taste about us. Rather he awaits us every instant in our action, in the work of the moment. There is a sense in which he is at the tip of my pen, my spade, my brush, my needle—of my heart and of my thought. By pressing the stroke, the line, or the stitch, on which I am engaged, to its ultimate natural finish, I shall lay hold of that last end towards which my innermost will tends. Like those formidable physical forces which man contrives to discipline so as to make them perform operations of prodigious delicacy, so the tremendous power of the divine attraction is focused on our frail desires and microscopic intents without breaking their point. It sur-animates; hence it neither disturbs anything nor stifles anything. It sur-animates; hence it introduces a higher principle of unity into our spiritual life, the specific effect of which is—depending upon the point of view one adopts— either to make man's endeavour holy, or to give the christian life the full flavour of humanity.

A. *The sanctification of human endeavour*

I do not think I am exaggerating when I say that nine out of ten practising Christians feel that man's work is always at the level of a "spiritual encumbrance." In spite of the practice of right intentions, and the day offered every morning to God, the general run of the faithful dimly feel that time spent at the office or the studio, in the fields or in the factory, is time taken away from prayer and adoration. It is impossible not to work—that is taken for granted. Then it is impossible, too, to aim at the deep religious life reserved for those who have the leisure to pray or preach all day long. A few moments of the day can be salvaged for God, yes, but the best hours are absorbed, or at any rate cheapened, by material cares. Under the sway of this feeling, large numbers of Catholics lead a double or crippled life in practice: they have to step out of their human dress so as to have faith in themselves as Christians—and inferior Christians at that.

What has been said above of the divine extensions and God-given demands of the mystical or universal Christ, should be enough to demonstrate both the emptiness of these impressions and the validity of the thesis (so dear to Christianity) of sanctification through fulfilling the duties of our station. There are, of course, certain noble and cherished moments of the day—those when we pray or receive the sacraments. Were it not for these moments of more efficient or explicit commerce with God, the tide of the divine omnipresence, and our perception of it, would weaken until all that was best in our human endeavour, without being entirely lost to the world, would be for us emptied of God. But once we have jealously safeguarded our relation to God encountered, if I may dare use the expression, "in his pure state" (that is to say in a state of being distinct from all the constituents of the world), there is no need to fear that the most trivial or the most absorbing of occupations should force us to depart from him. To repeat: by virtue of the Creation and, still more, of the Incarnation, *nothing* here below *is profane* for those who know how to see. On the contrary, everything is sacred to the men who can distinguish that portion of chosen being which is subject to Christ's drawing power in the process of consummation. Try, with God's help, to perceive the connection—even physical and natural—which binds your labour with the building of the kingdom of heaven; try to realise that heaven itself smiles upon you and, through your works, draws you to itself; then, as you leave church for the noisy streets, you will remain with only one feeling, that of continuing to immerse yourself in God. If your work is dull or exhausting, take refuge in the inexhaustible and becalming interest of progressing in the divine life. If your work enthrals you, then allow the spiritual impulse which matter communicates to you to enter into your taste for God whom you know better and desire more under the veil of his works. Never, at any time, "whether eating or drinking," consent to do anything without first of all realising its significance and constructive value *in Christo Jesu,* and pursuing it with all your might. This is not simply a commonplace precept for salvation: it is the very path to sanctity for each man according to his state and calling. For what is sanctity in a creature if not to adhere to God with the maximum of his strength?—and what does that maximum adherence to God mean if not the fulfilment—in the world organised around Christ—of the exact function, be it lowly or eminent, to which that creature is destined both by natural endowment and by supernatural gift?

Within the Church we observe all sorts of groups whose members are vowed to the perfect practice of this or that particular virtue: mercy, detachment, the splendour of the liturgy, the missions, contemplation. Why should there not be men vowed to the task of exemplifying, by their lives, the general sanctification of human endeavour?—men whose common religious ideal would be to give a full and conscious explanation of the divine possibilities or demands which any worldly occupation implies—men, in a word, who would devote themselves, in the fields of thought, art, industry, commerce and politics, etc., to carrying out in the sublime spirit these demands—the basic tasks which form the very bonework of human society? Around us the "natural" progress which nourishes the sanctity of each new age is all too often left to the children of the world, that is to say to agnostics or the irreligious. Unconsciously or involuntarily such men collaborate in the kingdom of God and in the fulfilment of the elect: their efforts, going beyond or correcting their incomplete or bad intentions, are gathered in by him "whose energy subjects all things to itself." But that is no more than a second best, a temporary phase in the organisation of human activity. Right from the hands that knead the dough, to those that consecrate it, the great and universal Host should be prepared and handled in a spirit of *adoration*.

May the time come when men, having been awakened

to a sense of the close bond linking all the movements of this world in the single, all-embracing work of the Incarnation, shall be unable to give themselves to any one of their tasks without illuminating it with the clear vision that their work—however elementary it may be—is received and put to good use by a Centre of the universe.

When that comes to pass, there will be little to separate life in the cloister from the life of the world. And only then will the action of the children of heaven (at the same time as the action of the children of the world) have attained the intended plenitude of its humanity.

B. *The humanisation of christian endeavour*

The great objection brought against Christianity in our time, and the real source of the distrust which insulates entire blocks of humanity from the influence of the Church, has nothing to do with historical or theological difficulties. It is the suspicion that our religion makes its followers *inhuman*.

"Christianity," so some of the best of the Gentiles are inclined to think, "is bad or inferior because it does not lead its followers to levels of attainment beyond ordinary human powers; rather it withdraws them from the ordinary ways of humankind and sets them on other paths. It isolates them instead of merging them with the mass. Instead of harnessing them to the common task, it causes them to lose interest in it. Hence, far from raising them to a higher level, it diminishes them and makes them false to their nature. Moreover, don't they admit as much themselves? And if one of their religious, or one of their priests, should happen to devote his life to research in one of the so-called secular disciplines, he is very careful, as a rule, to point out that he is only lending himself for a time to serve a passing whim of scholarly fashion or even something ultimately of the stuff of illusion, and that simply in order to show that Christians are not, after all, the stupidest of men. When a Catholic works with us, we invariably get the impression that he is doing so in a spirit of condescension. He appears to be interested, but in fact, because of his religion, he simply does not believe in the human effort as such. His heart is not really with us. Christianity nourishes deserters and false friends: that is what we cannot forgive."

We have placed this objection, which would be deadly if it were true, in the mouth of an unbeliever. But has it no echo, here and there, within the most faithful souls? What Christian who has become aware of a sheet of glass insulating him from his non-believing colleagues, has not asked himself uneasily whether he was not on a false tack or had not actually lost touch with the main current of mankind?

Without denying that some Christians, by their words more than their deeds, do give grounds for the reproach of being, if not the "enemies," at least the "stragglers" of the human race, we can safely assert, after what we said above concerning the supernatural value of our work on earth, that their attitude is due to an incomplete understanding and not at all to some ineradicable flaw in our religion.

How could we be deserters, or sceptical about the future of the tangible world? How could we be repelled by human labour? How little you know us! You suspect us of not sharing your concern and your hopes and your excitement as you penetrate the mysteries and conquer the forces of nature. "Feelings of this kind," you say, "can only be shared by men struggling side by side for existence; whereas you Christians profess to be saved already." As though for us as for you, indeed far more than for you, it were not a matter of life and death that the earth should flourish to the uttermost of its natural powers. As far as you are concerned (and it is here that you are not yet human enough, you do not *go to the limits* of your humanity) it is simply a matter of the success or failure of a reality which remains vague and precarious even when conceived in the form of some super-humanity. For us it is a question in a true sense of achieving the victory of no less than a God. One thing is infinitely disappointing, I grant you: far too many Christians are insufficiently conscious of the "divine" responsibilities of their lives, and live like other men, giving only half of themselves, never experiencing the spur or the intoxication of advancing God's kingdom in every domain of mankind. But do not blame anything but our weakness: our faith imposes on us the right and the duty to throw ourselves into the things of the earth. As much as you, and even better than you (because, of the two of us, I alone am in a position to prolong the perspectives of my endeavour to infinity, in conformity with the requirements of my present intention), I want to dedicate myself body and soul to the sacred duty of research. We must test every barrier, try every path, plumb every abyss. *Nihil intentatum* . . . God wills it, who willed that he should have need of it. You are men, you say? *Plus et ego.*

Plus et ego. There can be no doubt of it. At a time when the consciousness of its own powers and possibilities is legitimately awakening in a mankind now ready to become adult, one of the first duties of a Christian as an apologist is to show, by the logic of his religious views and still more by the logic of his action, that the incarnate God did not come to diminish in us the glorious responsibility and splendid ambition that is ours: *of fashioning our own self.* Once again, *non minuit, sed sacravit.* No, Christianity is not, as it is sometimes presented and sometimes practised, an additional burden of observances and obligations to weigh down and increase the already heavy load, or to multiply the already paralysing ties of our life in society. It is, in fact, a soul of immense power which bestows significance and beauty and a new lightness on what we are already doing. It is true that it sets us on the road towards unsuspected heights. But the slope which leads to these heights is linked so closely with the one we were already climbing naturally, that there is nothing so distinctively human in the Christian (and this is what remains to be considered) as his detachment.

6. DETACHMENT THROUGH ACTION

There hardly seems room for any dispute between Christians about what we have so far said about the *intrinsic* divinisation of human endeavour, since we have confined ourselves, in establishing it, to taking, in their proper strict sense, certain universally recognised theoretical and practical truths and confronting them with each other.

Nevertheless, some readers, though without finding any specific flaw in our argument, may feel vaguely upset or uneasy in the face of a christian ideal which lays such stress on the preoccupations of human development and the pursuit of earthly improvements. They should bear in mind that we are still only halfway along the road which leads to the mountain of the Transfiguration. Up to this point we have been dealing only with the active part of our lives. In a moment or two, when we come to the chapter on passivities and diminishment, the arms of the Cross will begin to dominate the scene more widely. Let us consider it for a moment. In the very optimistic and very broadening attitude which has been roughly sketched above, a true and deep renunciation lies concealed. Anyone who devotes himself to human duty according to the christian formula, though outwardly he may seem to be immersed in the concerns of the earth, is in fact, down to the depths of his being, a man of great detachment.

Of its very nature work is a manifold instrument of detachment, provided a man gives himself to it faithfully and without rebellion. In the first place it implies effort and a victory over inertia. And then, however interesting and intellectual it may be (and the more intellectual it is, the truer this becomes), work is always accompanied by the painful pangs of birth. Men can only escape the terrible boredom of monotonous and commonplace duty to find themselves a prey to the inner tension and the anxieties of "creation." To create, or organise, material energy, or truth, or beauty, brings with it an inner torment which prevents those who face its hazards from sinking into the quiet and closed-in life wherein grows the vice of self-regard and attachment (in the technical sense). An honest workman not only surrenders his calm and peace once and for all, but must learn continually to jettison the form which his labour or art or thought first took, and go in search of new forms. To pause, so as to bask in or possess results, would be a betrayal of action. Over and over again he must go beyond himself, tear himself away from himself, leaving behind him his most cherished beginnings. And on that road, which is not so different from the royal road of the Cross as might appear at first sight, detachment does not consist only in continually replacing one object with another of the same order—as miles, on a flat road, replace miles. By virtue of a marvellous mounting force contained in things (and which will be analysed in greater detail when we consider the "spiritual power of matter"), each reality attained and left behind gives us access to the discovery and pursuit of an ideal of higher spiritual content. Those who spread their sails in the right way to the winds of the earth will always find themselves borne by a current towards the open seas. The more nobly a man wills and acts, the more avid he becomes for great and sublime aims to pursue. He will no longer be content with family, country and the remunerative aspect of his work. He will want wider organisations to create, new paths to blaze, causes to uphold, truths to discover, an ideal to cherish and defend. So, gradually, the worker no longer belongs to himself. Little by little the great breath of the universe has insinuated itself into him through the fissure of his humble but faithful action, has broadened him, raised him up, borne him on.

It is in the Christian, provided he knows how to make the most of the resources of his faith, that these effects will reach their climax and their crown. As we have seen: from the point of view of the reality, accuracy and splendour of the ultimate end towards which we must aim in the least of our acts, we, disciples of Christ, are the most favoured of men. The Christian knows that his function is to divinise the world in Jesus Christ. In him, therefore, the natural process which drives human action from ideal to ideal and towards objects ever more internally coherent and comprehensive in their embrace, reaches—thanks to the support of Revelation—its fullest expansion. And in him, consequently, detachment through action should produce its maximum effectiveness.

And this is perfectly true. The Christian as we have described him in these pages, is at once the most attached and the most detached of men. Convinced in a way in which the "worldly" cannot be of the unfathomable importance and value concealed beneath the humblest worldly successes, the Christian is at the same time as convinced as the hermit of the worthlessness of any success which is envisaged only as a benefit to himself (or even a general one) without reference to God. It is God and God alone whom he pursues through the reality of created things. For him, interest lies truly *in* things, but in absolute dependence upon God's presence in them. The light of heaven becomes perceptible and attainable to him in the crystalline transparency of beings. But he wants only this light, and if the light is extinguished, whether because the object is out of its true place, or has outlived its function, or has moved itself, then even the most precious substance is only ashes in his sight. Similarly, within himself and his most personal development, it is not himself that he is seeking, but that which is greater than he, to which he knows that he is destined. In his own view he himself no longer counts, no longer exists; he has forgotten and lost himself in the very endeavour which is making him perfect. It is no longer the atom which lives, but the universe within it.

Not only has he encountered God in the entire field of his actions in the perceptible world, but in the course of this first phase of his spiritual development, the divine *milieu* which has been uncovered absorbs his powers in the very proportion in which these laboriously rise above their individuality.

PART TWO

⊚

The Divinisation of Our Passivities

While man by the very development of his powers is led to discover ever vaster and higher aims for his action, he also tends to be dominated by the object of his conquests and, like Jacob wrestling with the Angel, he ends by adoring what he was struggling against. The scale of that which he has unveiled and unleashed brings him into subjection. And then, because of his nature as element, he is brought to recognise that, in the final act that is to unite him to the All, the two terms of the union are utterly disproportionate. He, the lesser, has to receive rather than to give. He finds himself in the grip of what he thought he could grasp.

The Christian, who is by right the first and most human of men, is more subject than others to this psychological reversal whereby, in the case of all intelligent creatures, joy in action imperceptibly melts into desire for submission, and the exaltation of becoming one's own self into the zeal to die in another. Having been perhaps primarily alive to the attractions of union with God through action, he begins to conceive and then to desire a complementary aspect, an ulterior phase, in his communion: one in which he would not develop himself so much as lose himself in God.

He does not have to look far to discover possibilities and

opportunities for fulfilment in this gift of self. They are offered him at every moment—indeed they besiege him on all sides in the length and depth of the countless servitudes which make us servants far more than masters of the universe.

The moment has come to examine the number, the nature and the possible divinisation, of our passivities.

1. THE EXTENT, DEPTH AND DIVERSE FORMS OF HUMAN PASSIVITIES

The passivities of our lives, as we said at the beginning of this study, form half of human existence. The term means, quite simply, that that which is not done by us, is, by definition, undergone.

But this does not in any way prejudge the proportions in which action and passion possess our inner realm. In fact, these two parts of our lives—the active and the passive—are extraordinarily unequal. Seen from our point of view, the active occupies first place because we prefer it and because it is more easily perceived. But in the reality of things the passive is immeasurably the wider and the deeper part.

In the first place the passivities ceaselessly accompany our conscious deeds, in the form of reactions which direct, sustain or oppose our efforts. On this ground alone they inevitably and precisely coincide with the scope of our activities. But their sphere of influence extends far beyond these narrow limits. If we consider the matter carefully we in fact perceive with a sort of dismay that it is only the fine point of ourselves that comes up into the light of self-consciousness and freedom. We know ourselves and set our own course but within an incredibly small radius of light. Immediately beyond lies impenetrable darkness, though it is full of presences—the night of everything that is within us and around us, without us and in spite of us. In this darkness, as vast, rich, troubled and complex as the past and the present of the universe, we are not inert; we react, because we undergo. But this reaction, which operates without our control by an unknown prolongation of our being, is, humanly speaking, still a part of *our* passivity. In fact, everything beyond a certain distance is dark, and yet everything is full of being around us. This is the darkness, heavy with promises and threats, which the Christian will have to illuminate and animate with the divine presence.

In the midst of the confused energies which people this restless night, our mere presence immediately brings about the formation of two groups which press in upon us and demand to be treated in very different ways. On one side, the friendly and favourable forces, those which uphold our endeavour and point the way to success—the "passivities of growth." On the other side, the hostile powers which laboriously obstruct our tendencies, hamper or deflect our progress towards heightened being, and thwart our real or apparent capacities for development: these are the "passivities of diminishment."

Let us look at each group in turn; let us look them in the face until, in the depth of their alluring, unrevealing or hostile gaze, we discern the kindling light of the blessed countenance of God.

2. THE PASSIVITIES OF GROWTH AND THE TWO HANDS OF GOD

Growth seems so natural to us that we do not, as a matter of fact, pause to separate from our action the forces which nourish that action or the circumstances which favour its success.

And yet *quid habes quod non accepisti?* (what dost thou possess that thou hast not previously received?) We undergo life as much as we undergo death, if not more.

We must try to penetrate our most secret self, and examine our being from all sides. Let us try, patiently, to perceive the ocean of forces to which we are subjected and in which our growth is, as it were, steeped. This is a salutary exercise; for the depth and universality of our dependence on so much altogether outside our control all go to make up the embracing intimacy of our communion with the world to which we belong.

. . . And so, for the first time in my life perhaps (although I am supposed to meditate every day!), I took the lamp and, leaving the zone of everyday occupations and relationships where everything seems clear, I went down into my inmost self, to the deep abyss whence I feel dimly that my power of action emanates. But as I moved further and further away from the conventional certainties by which social life is superficially illuminated, I became aware that I was losing contact with myself. At each step of the descent a new person was disclosed within me of whose name I was no longer sure, and who no longer obeyed me. And when I had to stop my exploration because the path faded from beneath my steps, I found a bottomless abyss at my feet, and out of it came—arising I know not from where—the current which I dare to call *my* life.

What science will ever be able to reveal to man the origin, nature and character of that conscious power to will and to love which constitutes his life? It is certainly not our effort, nor the effort of anyone around us, which set that current in motion. And it is certainly not our anxious care, nor that of any friend of ours, which prevents its ebb or controls its turbulence. We can, of course, trace back through generations some of the antecedents of the torrent which bears us along; and we can, by means of certain moral and physical disciplines and stimulants, regularise or enlarge the aperture through which the torrent is released into us. But neither that geography nor those artifices help us in theory or in practice to harness the sources of life. My self is given to me far more than it is formed by me. Man, Scripture says, cannot add a cubit to his stature. Still less can he add a unit to the potential of his love, or accelerate by another unit the fundamental rhythm which regulates the ripening of his mind and heart. In the last resort the profound life, the fontal life, the newborn life, escape our grasp entirely.

Stirred by my discovery, I then wanted to return to the light of day and forget the disturbing enigma in the comfortable surroundings of familiar things—to begin living again at the surface without imprudently plumbing the depths of the abyss. But then, beneath this very spectacle of the turmoil of life, there reappeared, before my newly-opened eyes, the unknown that I wanted to escape. This time it was not hiding at the bottom of an abyss; it disguised its presence in the innumerable strands which form the web of chance, the very stuff of which the universe and my own small individuality are woven. Yet it was the same mystery without a doubt: I recognised it. Our mind is disturbed when we try to plumb the depth of the world beneath us. But it reels still more when we try to number the favourable chances which must coincide at every moment if the least of living things is to survive and to succeed in its enterprises. After the consciousness of being something other and something greater than myself—a second thing made me dizzy: namely, the supreme improbability, the tremendous unlikelihood of finding myself existing in the heart of a world that has survived and succeeded in being a world.

At that moment, as anyone else will find who cares to make this same interior experiment, I felt the distress characteristic to a particle adrift in the universe, the distress which makes human wills founder daily under the crushing number of living things and of stars. And if something saved me, it was hearing the voice of the Gospel, guaranteed by divine successes, speaking to me from the depth of the night: *ego sum, noli timere* (It is I, be not afraid).

Yes, O God, I believe it: and I believe it all the more willingly because it is not only a question of my being consoled, but of my being completed: it is you who are at the origin of the impulse, and at the end of that continuing pull which all my life long I can do no other than follow, or favour the first impulse and its developments. And it is you who vivify, for me, with your omnipresence (even more than my spirit vivifies the matter which it animates), the myriad influences of which I am the constant object. In the life which wells up in me and in the matter which sustains me, I find much more than your gifts. It is you yourself whom I find, you who makes me participate in your being, you who moulds me. Truly in the ruling and in the first disciplining of my living strength, in the continually beneficent play of secondary causes, I touch, as near as possible, the two faces of your creative action, and I encounter, and kiss, your two marvellous hands—the one which holds us so firmly that it is merged, in us, with the sources of life, and the other whose embrace is so wide that, at its slightest pressure, all the springs of the universe respond harmoniously together. By their very nature, these blessed passivities which are, for me, the will to be, the wish to be thus and thus, and the chance of fulfilling myself according to my desire, are all charged with your influence—an influence which will shortly appear more distinctly to me as the organising energy of the mystical body. In order to communicate with you in them in a fontal communion (a com-

munion in the sources of Life), I have only to recognise you in them, and to ask you to be ever more present in them.

O God, whose call precedes the very first of our movements, grant me the desire to desire being—that, by means of that divine thirst which is your gift, the access to the great waters may open wide within me. Do not deprive me of the sacred taste for being, that primordial energy, that very first of our points of rest: Spiritu principali confirma me. *And you whose loving wisdom forms me out of all the forces and all the hazards of the earth, grant that I may begin to sketch the outline of a gesture whose full power will only be revealed to me in presence of the forces of diminishment and death; grant that, after having desired, I may believe, and believe ardently and above all things, in your active presence.*

Thanks to you, that expectation and that faith are already full of operative virtue. But how am I to set about showing you and proving to myself, through some external effort, that I am not one of those who say Lord, Lord! With their lips only? I shall work together with your action which ever forestalls me, and will do so doubly. First, to your deep inspiration which commands me to be, I shall respond by taking great care never to stifle nor distort nor waste my power to love and to do. Next, to your all-embracing providence which shows me at each moment, by the day's events, the next step to take and the next rung to climb, I shall respond by my care never to miss an opportunity of rising "towards the level of spirit."

The life of each one of us is, as it were, woven of those two threads: the thread of inward development, through which our ideas and affections and our human and religious attitudes are gradually formed; and the thread of outward success by which we always find ourselves at the exact point where the whole sum of the forces of the universe meet together to work in us the effect which God desires.

O God, that at all times you may find me as you desire me and where you would have me be, that you may lay hold on me fully, both by the Within and the Without of myself, grant that I may never break this double thread of my life.

3. THE PASSIVITIES OF DIMINISHMENT*

To cleave to God hidden beneath the inward and outward forces which animate our being and sustain it in its development, is ultimately to open ourselves to, and put trust in, all the breaths of life. We answer to, and "communicate" with, the passivities of growth by our fidelity in action. Hence by our very desire to experience God passively we find ourselves brought back to the lovable duty of growth.

The moment has come to plumb the decidedly negative side of our existences—the side on which, however far we search, we cannot discern any happy result or any solid conclusion to what happens to us. It is easy enough to understand that God can be grasped in and through every life. But can God also be found in and through every death? This is what perplexes us deeply. And yet this is what we must learn to acknowledge as a matter of settled habit and practice, unless we abandon all that is most characteristically christian

---

*If, in speaking of evil in this section, we do not mention sin more explicitly, it is because the aim of the following pages being solely to show how all things can help the believer to unite himself to God, there is no need to concern ourselves directly with bad actions, that is, with positive gestures of disunion. Sin only interests us here in so far as it is a weakening, a deviation caused by our personal faults (even when repented), or the pain and the scandal which the faults of others inflict on us. From this point of view it makes us suffer and can be transformed in the same way as any other suffering. That is why physical evil and moral evil are presented here, almost without distinction, in the same chapter on the passivities of diminishment.

in the christian outlook; and unless we are prepared to forfeit commerce with God in one of the most widespread and at the same time most profoundly passive and receptive experiences of human life.

The forces of diminishment are our real passivities. Their number is vast, their forms infinitely varied, their influence constant. In order to clarify our ideas and direct our meditation we will divide them into two groups corresponding to the two forms under which we considered the forces of growth: the diminishments whose origin lies *within us,* and the diminishments whose origin lies *outside us.*

The external passivities of diminishment are all our bits of ill fortune. We have only to look back on our lives to see them springing up on all sides: the barrier which blocks our way, the wall that hems us in, the stone which throws us from our path, the obstacle that breaks us, the invisible microbe that kills the body, the little word that infects the mind, all the incidents and accidents of varying importance and varying kinds, the tragic interferences (upsets, shocks, severances, deaths) which come between the world of "other" things and the world that radiates out from us. And yet when hail, fire and thieves had taken everything from Job—all his wealth and all his family—Satan could say to God: "Skin for skin, and all that a man hath he will give for his life. But put forth thy hand, and touch his bone and his flesh: and then thou shalt see that he will bless thee to thy face." In a sense the loss of things means little to us because we can always imagine getting them back. What is terrible for us is to be cut off from things through some inward diminishment that can never be retrieved.

Humanly speaking, the internal passivities of diminishment form the darkest element and the most despairingly useless years of our life. Some were waiting to pounce on us as we first awoke: natural failings, physical defects, intellectual or moral weaknesses, as a result of which the field of our activities, of our enjoyment, of our vision, has been pitilessly limited since birth. Others were lying in wait for us later on and appeared as suddenly and brutally as an accident, or as stealthily as an illness. All of us one day or another will come to realise, if we have not already done so, that one or other of these sources of disintegration has lodged itself in the very heart of our lives. Sometimes it is the cells of the body that rebel or become diseased; at other times the very elements of our personality seem to be in conflict or to detach themselves from any sort of order. And then we impotently stand by and watch collapse, rebellion and inner tyranny, and no friendly influence can come to our help. And if by chance we escape, to a greater or lesser extent, the critical forms of these assaults from without which appear deep within us and irresistibly destroy the strength, the light and the love by which we live, there still remains that slow, essential deterioration which we cannot escape: old age little by little robbing us of ourselves and pushing us on towards the end. Time, which postpones possession, time which tears us away from enjoyment, time which condemns us all to death—what a formidable passivity is the passage of time. . . .

In death, as in an ocean, all our slow or swift diminishments flow out and merge. Death is the sum and consummation of all our diminishments: it is *evil* itself—purely physical evil, in so far as it results organically in the manifold structure of that physical nature in which we are immersed—but a moral evil too, in so far as in the society to which we belong, or in ourselves, the wrong use of our freedom, by spreading disorder, converts this manifold complexity of our nature into the source of all evil and all corruption.

We must overcome death by finding God in it. And by

the same token, we shall find the divine established in our innermost hearts, in the last stronghold which might have seemed able to escape his reach.

Here again, as in the case of the "divinisation" of our human activities, we shall find the christian faith absolutely explicit in what it claims to be the case, and what it bids us do. Christ has conquered death, not only by suppressing its evil effects, but by reversing its sting. By virtue of Christ's rising again, nothing any longer kills inevitably but everything is capable of becoming the blessed touch of the divine hands, the blessed influence of the will of God upon our lives. However marred by our faults, or however desperate in its circumstances, our position may be, we can, by a total re-ordering, completely correct the world that surrounds us, and resume our lives in a favourable sense. *Diligentibus Deum omnia convertuntur in bonum.* That is the fact which dominates all explanation and all discussion.

But here again, as in the matter of the saving value of our human endeavour, our mind wants to validate to itself its hopes so as to surrender to them more completely.

Quomodo fiet istud? This study is all the more necessary because the christian attitude to evil lends itself to some very dangerous misunderstandings. A false interpretation of christian resignation, together with a false idea of christian detachment, is the principal source of the antagonisms which make a great many Gentiles so sincerely hate the Gospel.

Let us ask ourselves how, and in what circumstances, our apparent deaths, that is to say the waste-matter of our existences, can find their necessary place in the establishment, around us, of the kingdom of God and the *milieu* of God. It will help us to do this if we thoughtfully distinguish two phases, two periods, in the process which culminates in the transfiguration of our diminishments. The first of these phases is that of our struggle against evil. The second is that of defeat and of its transfiguration.

A. *Our struggle with God against evil*

When a Christian suffers, he says "God has touched me." The words are pre-eminently true, though their simplicity summarises a very complex series of spiritual operations; and it is *only when we have gone right through that whole series of operations* that we have the right to speak those words. For if, in the course of our encounters with evil, we try to distinguish what the Schoolmen term "the instants of nature," we shall have, on the contrary, to begin by saying "God wants to free me from this diminishment—God wants me to help him to take this cup from me." To struggle against evil and to reduce to a minimum even the ordinary physical evil which threatens us, is unquestionably the first act of our Father who is in heaven; it would be impossible to conceive him in any other way, and still more impossible to love him.

It is a perfectly correct view of things—and strictly consonant with the Gospel—to regard Providence across the ages as brooding over the world in ceaseless effort to spare that world its bitter wounds and to bind up its hurts. Most certainly it is God himself who, in the course of the centuries, awakens the great benefactors of humankind, and the great physicians, in ways that agree with the general rhythm of progress. He it is who inspires, even among those furthest from acknowledging his existence, the quest for every means of comfort and every means of healing. Do not men acknowledge by instinct this divine presence when hatreds are quenched and their protesting uncertainty resolved as they kneel to thank each one of those who have helped their body or their mind to freedom? Can there be any doubt of it? At the first approach of the diminishments we cannot hope to find God except by loathing what is coming upon us and doing our best to avoid it. The more we repel suffering at that moment, with our whole heart and our whole strength,* the more closely we cleave to the heart and action of God.

B. *Our apparent failure and its transfiguration*

With God as our ally we are always certain of saving our souls. But we know too well that there is no guarantee that we shall always avoid suffering or even those inward defeats on account of which we can imagine our lives to ourselves as failures. In any event, all of us are growing old and all of us will die. This means to say that, however fine our resistance, at some moment or other we feel the constraining grip of the forces of diminishment, against which we were fighting, gradually gaining mastery over the forces of life, and dragging us, physically vanquished, to the ground. But how can we be defeated if God is fighting on our side? or what does this defeat mean?

The problem of evil, that is to say the reconciling of our failures, even the purely physical ones, with creative goodness and creative power, will always remain one of the most disturbing mysteries of the universe for both our hearts and our minds. A full understanding of the suffering of God's creatures (like that of the pains of the damned) presupposes in us

*Without bitterness and without revolt, of course, but with an *anticipatory tendency* to acceptance and final resignation. It is obviously difficult to separate the two "instants of nature" without to some extent distorting them in describing them. But there is this to note: the necessity of the initial stage of resistance to evil is clear, and everyone admits it. The failure that follows on laziness, the illness contracted as a result of unjustified imprudence, could not be regarded by anyone as being the *immediate* will of God.

an appreciation of the nature and value of "participated being" which, for lack of any point of comparison, we cannot have. Yet this much we can see: on the one hand, the work which God has undertaken in uniting himself intimately to created beings presupposes in them a slow preparation in the course of which they *(who already exist, but are not yet complete)* cannot of their nature avoid the risks (increased by an original fault) involved in the imperfect ordering of the manifold, in them and around them; and on the other hand, because the final victory of good over evil can only be completed in the *total* organisation of the world, our infinitely short individual lives could not hope to know the joy, here below, of entry into the Promised Land. We are like soldiers who fall during the assault which leads to peace. God does not therefore suffer a preliminary defeat in our defeat because, although we appear to succumb individually, the world, in which we shall live again, triumphs in and through our deaths.

But this first aspect of his victory, which is enough to assure us of his omnipotence, is made complete by another disclosure—perhaps more direct and in every case more immediately experienceable by each of us—of his universal authority. In virtue of his very perfections,* God cannot ordain that the elements of a world in the course of growth—or at least of a fallen world in the process of rising again—should avoid shocks and diminishments, even moral ones: *necessarium est ut scandala eveniant*. But God will make it good—he will take his revenge, if one may use the expression—by making evil itself serve a higher good of his

---

*Because his perfections cannot run counter to the nature of things, and because a world, assumed to be progressing towards perfection, or "rising upward," is of its nature precisely still partially disorganised. A world without a trace or a threat of evil would be a world already consummated.

faithful, the very evil which the present state of creation does not allow him to suppress immediately. Like an artist who is able to make use of a fault or an impurity in the stone he is sculpting or the bronze he is casting so as to produce more exquisite lines or a more beautiful tone, God, without sparing us the partial deaths, nor the final death, which form an essential part of our lives, transfigures them by integrating them in a better plan—*provided we lovingly trust in him*. Not only our unavoidable ills but our faults, even our most deliberate ones, can be embraced in that transformation, provided always we repent of them. Not everything is immediately good to those who seek God; but everything is capable of becoming good: *omnia convertuntur in bonum*.*

What is the process and what are the phases by which God accomplishes this marvellous transformation of our deaths into a better life? Drawing on analogies from what we know how to bring about ourselves, and reflecting on the constant attitude and practical teaching of the Church with regard to human suffering, we may perhaps hazard an answer to this question.

It could be said that Providence, for those who believe in it, converts evil into good in three principal ways. Sometimes the check we have undergone will divert our activity on to objects, or towards a framework, that are more propitious—though still situated on the level of the human ends we are pursuing. That is what happened with Job, whose final happiness was greater than his first. At other times, more often perhaps, the loss which afflicts us will oblige us to turn for the satisfaction of our frustrated desires to less material fields, which neither worm nor rust can corrupt. The lives of

---

*On the "miraculous" effects of faith, see p. 111. There is obviously no intention of giving a general theory of prayer here.

the saints and, generally speaking, the lives of all those who have been outstanding for intelligence or goodness, are full of these instances in which one can see the man emerging ennobled, tempered and renewed from some ordeal, or even some downfall, which seemed bound to diminish or lay him low for ever. Failure in that case plays for us the part that the elevator plays for an aircraft or the pruning knife for a plant. It canalises the sap of our inward life, disengages the purest "components" of our being in such a way as to make us shoot up higher and straighter. The collapse, even when a moral one, is thus transformed into a success which, however spiritual it may be is, nevertheless, felt *experimentally*. In the presence of St. Augustine, St. Mary Magdalen or St. Lydwine, no one hesitates to think *felix dolor* or *felix culpa*. With the result that, up to this point, we still "understand" Providence.

But there are more difficult cases (the most common ones, in fact) where human wisdom is altogether out of its depth. At every moment we see diminishment, both in us and around us, which does not seem to be compensated by advantages on any perceptible plane: premature deaths, stupid accidents, weaknesses affecting the highest reaches of our being. Under blows such as these, man does not move upward in any direction that we can perceive; he disappears or remains grievously diminished. How can these diminishments which are altogether without compensation, wherein we see death at its most deathly, become for us a good? This is where we can see the third way in which Providence operates in the domain of our diminishments—the most effective way and the way which most surely makes us holy.

God, as we have seen, has already transfigured our sufferings by making them serve our conscious fulfilment. In his hands the forces of diminishment have perceptibly become the tool that cuts, carves and polishes within us the stone which is destined to occupy a definite place in the heavenly Jerusalem. But he will do still more, for, as a result of his omnipotence impinging upon our faith, events which show themselves experimentally in our lives as pure loss will become an immediate factor in the union we dream of establishing with him.

Uniting oneself means, in every case, migrating, and dying partially in what one loves. But if, as we are sure, this being reduced to nothing in the other must be all the more complete the more we give our attachment to one who is greater than ourselves, then we can set no limits to the tearing up of roots that is involved on our journey into God. The progressive breaking-down of our self-regard by the "automatic" broadening of our human perspectives (analysed above on pp. 36–37), when accompanied by the gradual spiritualisation of our tastes and aspirations under the impact of certain setbacks, is no doubt a very real foretaste of that leap out of ourselves which must in the end deliver us from the bondage of ourselves into the service of the divine sovereignty. Yet the effect of this initial detachment is for the moment only to develop the centre of our personality to its utmost limits. Arrived at that ultimate point we may still have the impression of possessing ourselves in a supreme degree—of being freer and more active than ever. We have not yet crossed the critical point of our ex-centration, of our reversion to God. There is a further step to take: the one that makes us *lose all foothold within ourselves—oportet illum crescere, me autem minui*. We are still not lost to ourselves. What will be the agent of that definitive transformation? Nothing else than death.

In itself, death is an incurable weakness of corporeal beings, complicated, in our world, by the influence of an original fall. It is the sum and type of all the forces that diminish us, and against which we must fight without being able to hope for a personal, direct and immediate victory. Now the great victory of the Creator and Redeemer, in the christian vision, is to have transformed what is in itself a universal power of diminishment and extinction into an essentially life-giving factor. God must, in some way or other, make room for himself, hollowing us out and emptying us, if he is finally to penetrate into us. And in order to assimilate us in him, he must break the molecules of our being so as to re-cast and re-model us. The function of death is to provide the necessary entrance into our inmost selves. It will make us undergo the required dissociation. It will put us into the state organically needed if the divine fire is to descend upon us. And in that way its fatal power to decompose and dissolve will be harnessed to the most sublime operations of life. What was by nature empty and void, a return to bits and pieces, can, in any human existence, become fullness and unity in God.

C. *Communion through diminishment*

It was a joy to me, O God, in the midst of the struggle, to feel that in developing myself I was increasing the hold that you have upon me; it was a joy to me, too, under the inward thrust of life or amid the favourable play of events, to abandon myself to your providence. Now that I have found the joy of utilising all forms of growth to make you, or to let you, grow in me, grant that I may willingly consent to this last phase of communion in the course of which I shall possess you by diminishing in you.

After having perceived you as he who is "a greater myself," grant, when my hour comes, that I may recognize you under the species of each alien or hostile force that seems bent upon destroying or uprooting me. When the signs of age begin to mark my body (and still more when they touch my mind); when the ill that is to diminish me or carry me off strikes from without or is born within me; when the painful moment comes in which I suddenly awaken to the fact that I am ill or growing old; and above all at that last moment when I feel I am losing hold of myself and am absolutely passive within the hands of the great unknown forces that have formed me; in all those dark moments, O God, grant that I may understand that it is you (provided only my faith is strong enough) who are painfully parting the fibres of my being in order to penetrate to the very marrow of my substance and bear me away within yourself.

The more deeply and incurably the evil is encrusted in my flesh, the more it will be you that I am harbouring—you as a loving, active principle of purification and detachment. The more the future opens before me like some dizzy abyss or dark tunnel, the more confident I may be—if I venture forward on the strength of your word—of losing myself and surrendering myself in you, of being assimilated by your body, Jesus.

You are the irresistible and vivifying force, O Lord, and because yours is the energy, because, of the two of us, you are infinitely the stronger, it is on you that falls the part of consuming me in the union that should weld us together. Vouchsafe, therefore, something more precious still than the grace for which all the faithful pray. It is not enough that I should die while communicating. Teach me to treat my death as an act of communion.

D. *True resignation*

The above analysis (in which we have tried to distinguish the phases by which our diminishments may be divinised) has helped us to *validate to ourselves* the christian formula, which is so comforting to those who suffer, "God has touched me. God has taken away from me. His will be done." As a result of this analysis we have understood how the two hands of God can reappear, more active and more penetrating than ever, beneath the evils that corrupt us from within, and the blows that break us up from without. But the analysis has a further result, almost as priceless as the first. It puts those of us who are Christians in a position to justify to those who are not Christians the legitimacy and the human value of resignation.

There are many reasonable men who honestly consider and denounce christian resignation as being one of the most dangerous and soporific elements in "the opium of the people." Next to disgust with the earth, there is no attitude which the Gospel is so bitterly reproached with having fostered as that of passivity in the face of evil—a passivity which can go as far as a perverse cultivation of suffering and diminishment. As we have already said, with reference to "false detachment": this accusation, or even suspicion, is infinitely more effective, at this moment, in preventing the conversion of the world than all the objections drawn from science or philosophy. A religion which is judged to be inferior to our human ideal—in spite of the marvels by which it is surrounded—is already *condemned*. It is therefore of supreme importance for the Christian to understand and live submission to the will of God in the *active* sense which, as we have said, is the only orthodox sense.

No, if he is to practise to the full the perfection of his Christianity, the Christian must not falter in his duty to resist evil. On the contrary, during the first phase, as we have seen, he must fight sincerely and with all his strength, in union with the creative force of the world, to drive back evil—so that nothing in him or around him may be diminished. During this initial phase, the believer is the convinced ally of all those who think that humanity will not succeed unless it strives with all its might to realise its potentialities. And as we said with reference to human development, the believer is more closely tied than anyone to this great task, because in his eyes the victory of humanity over the diminishments of the world—even physical and natural—to some extent conditions the fulfilment and consummation of the quite specific Reality which he adores. As long as resistance is possible, the son of heaven will resist too—as firmly as the most worldly children of the world—everything that deserves to be scattered or destroyed.

Should he meet with defeat—the personal defeat which no human being can hope to escape in his brief single combat with forces whose order of magnitude and evolution are universal—he will, like the conquered pagan hero, still inwardly resist. Though he is stifled and constrained, his efforts will still be sustained. At that point, however, he will see a new realm of possibilities open out before him, instead of having nothing to compensate for and master his coming death except the melancholy and questionable consolation of stoicism (which, if carefully analysed, would probably prove in the end to owe its beauty and consistency to a despairing faith *in the value of sacrifice*). This hostile force that lays him low and disintegrates him can become for him a loving principle of renewal, if he accepts it with faith while never ceasing to struggle against it. On the experimental plane, everything is lost. But in the realm of the supernatural, as it is called, *there is a further dimension* which allows God to achieve, *insensibly,* a mysterious reversal of evil into good. Leaving the zone of human successes and failures behind him, the Christian accedes by an effort of trust in the greater than himself to the region of suprasensible transformations and growth. His resignation is no more than the thrust which lifts the field of his activity higher.

We have come a long way, christianly speaking, from the justly criticised notion of "submission to the will of God" which is in danger of weakening and softening the fine steel of the human will, brandished against all the powers of darkness and diminishment. We must understand this well and cause it to be understood: to find and to do the will of God (even as we diminish and as we die) does not imply either a direct encounter or a passive attitude. I have no right to regard the evil that comes upon me through my own negligence or fault as being the touch of God.* I can only unite myself to the will of God (as endured passively) *when all my strength is spent,* at the point where my activity, fully extended and straining towards betterment (understood in ordinary human terms), finds itself continually counter-weighted by forces tending to halt me or overwhelm me. Unless I do everything I can to advance or resist, I shall not find myself at the *required point*—I shall not submit to God as much as I might have done or as much as he wishes. If, on the contrary, I persevere courageously, I shall rejoin God across evil, deeper down than evil; I shall draw close to him; and at that moment the optimum of my "communion in resignation" necessarily coincides (by definition) with the maximum of fidelity to the human task.

FRENCH EDITOR'S NOTE

It is interesting to compare these pages on "the divinisation of the activities and passivities" with the following clarifications taken from a

---

*Though the harm which results from my negligence can become the will of God for me on condition I repent and correct my lazy or indifferent attitude of mind. Everything can be taken up again and recast in God, even one's faults.

letter written shortly before *Le Milieu Divin,* in which the author sets out his spiritual doctrine to Père Auguste Valensin, one of his closest friends:

I agree, fundamentally, that the completion of the world is only consummated through a death, a "night," a reversal, an ex-centration, and a quasi-depersonalisation. . . . Union with Christ presupposes essentially that we transpose the ultimate centre of our existence into him—which implies the radical sacrifice of egoism. . . .

(Nevertheless)

If Christ is to take possession of all my life—of all life—then it is essential that I should grow in him not only by means of the ascetic constraints and the supremely unifying severances of suffering, but also by means of everything that my existence brings with it of positive effort, and the perfecting of my nature.

The formula for renunciation, if it is to be total, must satisfy two conditions:

1. it must enable us to go beyond everything there is in the world
2. and yet at the same time compel us to press forward (with conviction and passion) the development of this same world.

Speaking in general, Christ gives himself to us through the world which is to be consummated in relation to him.

You should note the following point carefully: I do not attribute any definitive or absolute value to the varied constructions of nature. What I like about them is not their particular form, but their function, which is to build up mysteriously, first what can be divinised, and then, through the grace of Christ coming down upon our endeavour, what is divine. . . .

To sum up, *complete* christian endeavour consists, in my view, in three things:

1. collaborating passionately in the human effort in the conviction that, not only through our fidelity and obedience, but

also through the *work* realised, we are working for the fulfilment of the Pleroma by preparing its more or less near-to-
our-hand material

2. in the course of this hard labour, and in the pursuit of an ever
widening ideal, achieving a preliminary form of renunciation
and of victory over a narrow and lazy egoism

3. cherishing the 'hollownesses' as well as the 'fullnesses' of
life—that is to say its passivities and the providential diminishments through which Christ transforms directly and emi-
nently into himself the elements and the personality which we
have tried to develop for him.

In that way detachment and human endeavour are harmonised.
It should be added that the ways in which they can be combined are
infinitely varied. There is an infinity of vocations. Within the Church
there are St. Thomas Aquinas and St. Vincent de Paul side by side
with St. John of the Cross. There is a time for growth and a time for
diminishment in the lives of each one of us. At one moment the dominant note is one of constructive human effort, and at another mystical
annihilation. . . .

All these attitudes spring from the same inner orientation of the
mind, from a single law which combines the twofold movement of the
natural personalisation of man and his supernatural depersonalisation
in Christo. . . .

CONCLUSION TO PARTS ONE AND TWO

Some General Remarks on Christian Asceticism

Having observed the progressive invasion of divinisation into the active and passive halves of our lives, we are now in a position to take a general view of the heavenly layers into which this tide of light has plunged us. That will form the third part of this work.

But before setting ourselves to contemplate the divine *milieu*, we must, for the sake of clarity, sum up in general terms the ascetic doctrine running through the preceding pages.

We shall do this in three sections under the following headings: 1. Attachment and detachment; 2. The meaning of the Cross; 3. The spiritual power of matter.

1. ATTACHMENT AND DETACHMENT

Nemo dat quod non habet. No sweet-smelling smoke without incense. No sacrifice without a victim. How would man give himself to God if he did not exist? What possession could he transfigure by his detachment if his hands were empty?

These common-sense observations enable us to solve, in principle, the question which is formulated often rather clumsily in the following terms: "Which is better for a

Christian: activity or passivity? Life or death? Growth or diminishment? Development or curtailment? Possession or renunciation?"

The general answer to this is: "Why separate and contrast the two natural phases of a single effort? Your essential duty and desire is to be united with God. But in order to be united, you must first of all *be*—be yourself as completely as possible. And so you must develop yourself and take possession of the world *in order to be*. Once this has been accomplished, then is the time to think about renunciation; then is the time to accept diminishment for the sake of *being in another*. Such is the sole and twofold precept of complete christian asceticism."

Let us consider the two terms of this method more closely, and observe their particular interplay and the resulting effect.

A. First, develop yourself,* Christianity says to the Christian

Books about the spiritual life do not generally throw this first phase of christian perfection into clear enough relief. Perhaps

---

*"First," in this sense, clearly indicates a priority in nature as much as, or more than, a priority in time. The true Christian should obviously never be *purely and simply* attached to whatever it may be, because the contact he seeks with things is always made *with a view to* transcending them or transfiguring them. So that when we speak here of being attached, we mean something penetrated and dominated by detachment. (See below in the text.)

Nevertheless, the use and proportion of *development* in the spiritual life are very delicate matters, for nothing is easier than to pursue one's selfish interests under cover of growing and of loving in God. The only real protection against that dangerous illusion is a constant concern to keep very much alive (with God's help) the impassioned vision of the *Greater than All*. In the presence of that supreme interest, the very idea of growing or enjoying egotistically, for oneself, becomes insipid and intolerable.

it seems too obvious to deserve mention, or seems to belong too completely to the "natural" sphere, or possibly it is too dangerous to be insisted upon—whatever the reason, these books usually remain silent on the subject or take it for granted. This is a fault and an omission. Although the majority of people understand it easily enough, and although its essentials are common to the ethics of both layman and religious, the duty of human perfection, like the whole universe, has been renewed, recast, supernaturalised, in the kingdom of God. It is a truly christian duty to grow, even in the eyes of men, and to make one's talents bear fruit, even though they be natural. It is part of the essentially Catholic vision to look upon the world as maturing—not only in each individual or in each nation, but in the whole human race—a specific power of knowing and loving whose transfigured term is charity, but whose roots and elemental sap lie in the discovery and the love of everything that is true and beautiful in creation. This has already been explained with reference to the christian value of action; but here is the place to recall it: the effort of mankind, even in realms inaccurately called profane, must, in the christian life, assume the role of a holy and unifying operation. It is the collaboration, trembling with love, which we give to the hands of God, concerned to attire and prepare us (and the world) for the final union through sacrifice. Understood in this way, the care which we devote to personal achievement and embellishment is no more than a gift begun. And that is why the attachment to creatures which it appears to denote melts imperceptibly into complete detachment.

B. *And if you possess something, Christ says in the Gospel, leave it and follow me*

Up to a certain point the believer who, understanding the christian meaning of development, has worked to mould himself and the world for God, will hardly need to hear the second injunction before beginning to obey it. Anyone whose aim, in conquering the earth, has really been to subject a little more matter to spirit has, surely, begun to take leave of himself at the same time as taking possession of himself. This is also true of the man who rejects mere enjoyment, the line of least resistance, the easy possession of things and ideas, and sets out courageously on the path of work, inward renewal and the ceaseless broadening and purification of his ideal. And it is true, again, of the man who has given his time, his health, or his life, to something greater than himself— a family to be supported, a country to be saved, a truth to be discovered, a cause to be defended. All these men are continually passing from attachment to detachment as they faithfully mount the ladder of human endeavour.

There are, however, two forms of renunciation which are reserved, and the Christian will not embark upon them except at the invitation or on the express order of his Creator. We refer to the practice of the evangelical counsels and the use of diminishments, neither of which is justified by the pursuit of a clearly defined higher good.

Where the first are concerned, no-one will deny that the religious life (which was also discovered, and is still practised, outside Christianity) can be a normal and "natural" flowering of human activity in search of a higher life. Nevertheless the practice of the virtues of poverty, chastity and obedience does represent the beginnings of a flight beyond the normal spheres of earthly, procreative and conquering humanity; and for this reason they had to wait, before becoming generally valid and licit, for a *Duc in Altum* to authenticate the aspirations maturing in the human soul. That authorisation was given once and for all in the Gospel by the Master of things.

But it must also be heard individually by those who are to benefit from it: it is "vocation."

With the practice of the forces of diminishment, the initiative must, even more clearly, come entirely from God. Man can and should make use of penances of some kind to organise the hierarchy of, and liberate, the lower powers within him. He can and should sacrifice himself when a greater interest claims him. But he has not the right to diminish himself for the sake of diminishing himself. Voluntary mutilation, even when conceived as a method of inward liberation, is a crime against being, and Christianity has always explicitly condemned it. The Church's most firmly established teaching is that it is our duty as creatures to try and live more and more by the higher parts of ourselves, in conformity with the aspirations of the present life. That alone is our concern. The rest belongs to the wisdom of him who alone can bring forth another life from every form of death.

There is no need to be wildly impatient. The Master of death will come soon enough—and perhaps we can already hear his footsteps. There is no need to forestall his hour nor to fear it. When he enters into us to destroy, as it seems, the virtues and the forces that we have distilled with so much loving care out of the sap of the world, it will be as a loving fire to consummate our completion in union.

C. *Thus, in the general rhythm of christian life, development and renunciation, attachment and detachment, are not mutually exclusive*

On the contrary, they harmonise, like breathing in and out in the movement of our lungs. They are two phases of the soul's breath, or two components of the impulse by which the christian life uses things as a springboard from which to mount beyond them.*

That is the general solution. In the detail of particular cases, the sequence of these two phases and the combinations of these two components are subject to an infinite number of subtle variations. Their exact blending calls for a spiritual tact which is the strength and virtue proper to the masters of the inner life. In some Christians detachment will always retain the form of disinterestedness and endeavour, which belongs to human work faithfully carried out: the transfiguration of life will be wholly inward. In others a physical or moral caesura will occur in the course of their lives which will cause them to pass from the level of a very holy normal life to the level of elected renunciations and mystical states. But for all of them, in any event, the road ends at the same point: the final stripping in death which accompanies the recasting, and is a prelude to the final incorporation, *in Christo Jesu*. And for all of them again, which makes or mars their life is the degree of harmony with which the two factors of growing for Christ, and diminishing in him, are combined in the light of the natural and supernatural aptitudes involved. It would clearly be as absurd to prescribe unlimited development or renunciation as it would be to set no bounds to eating or

---

*From this "dynamic" point of view the opposition so often stressed between asceticism and mysticism disappears. There is nothing in man's concern for self-perfection to distract him from his absorption in God, provided the ascetic effort is simply the beginning of "mystical annihilation." There is no longer any reason to distinguish between an (ascetic) "anthropomorphism" and a (mystical) "theocentrism" once the human centre is seen and loved in conjunction with (that is, in movement towards) the divine centre. Of course as God takes possession of man, the creature finally becomes passive (because it finds itself newly created in the divine union). But that passivity presupposes a subject that reacts and an active phase. The fire of heaven must come down on something: otherwise there would be nothing consumed and nothing consummated.

fasting. In the spiritual life, as in all organic processes, everyone has their *optimum* and it is just as harmful to go beyond it as not to attain it.*

What has been said of individuals must be transposed and applied to the Church as a whole. It is probable that the Church is led, at different times in the course of her existence, to emphasise in her general life now a greater care to collaborate in the earthly task, now a more jealous concern to stress the ultimate transcendence of her preoccupations. What is quite certain is that her health and integrity, at any given moment, depend upon the exactitude with which her members, each in their proper place, fulfil their functions which range from the duty of applying themselves to what are reputed to be the most profane of worldly occupations, to vocations which call for the most austere penances or the most sublime contemplation. All those different roles are necessary. The Church is like a great tree whose roots must be energetically anchored in the earth while its leaves are serenely exposed to the bright sunlight. In this way she sums up a whole gamut of beats in a single living and all-embracing act, each one of which corresponds to a particular degree or a possible form of spiritualisation.

*One thus evades the basic problem of the use of creatures if one solves it by saying that in all cases the *least possible* should be taken from them. This *minimum* theory is no doubt the product of the mistaken notion that God grows in us by destruction or substitution rather than by transformation (see note, p. 80) or, which comes to the same thing, that the spiritual potential of the material creation is now exhausted. The *minimum* theory may be useful in reducing certain seeming risks; it does not teach us how to get the maximum spiritual yield from the objects which surround us—which is what the reign of God really means. The one absolute rule upon which we can depend in this matter would seem to be this: "To love in the world, in God, something which may always become greater." All the rest is a matter of christian prudence and individual vocation. See pp. 75–77 on the utilisation by each of us of the spiritual forces in matter.

In the midst of all that diversity there is, however, something which dominates—something which confers its distinctively christian character on the organism as a whole (as well as upon each element in it): it is the impulse towards the heavens, the laborious and painful bursting out beyond matter. It is important to remember (and we have not finished insisting on it) that the supernatural awaits and sustains the progress of our nature. But it must not be forgotten that it purifies and perfects that progress, in the end, only in an apparent annihilation. The inseparable alliance between the two terms, personal progress and renunciation in God; but also the continual, and then final, ascendency of the second over the first—it is these that sum up the full meaning of the mystery of the Cross.

2. THE MEANING OF THE CROSS

The Cross has always been a symbol of conflict, and a principle of selection, among men. The Faith tells us that it is by the willed attraction or repulsion exercised upon souls by the Cross that the sorting of the good seed from the bad, the separation of the chosen elements from the unutilisable ones, is accomplished at the heart of mankind. Wherever the Cross appears, unrest and antagonisms are inevitable. But there is no reason why these conflicts should be needlessly exacerbated by preaching the doctrine of Christ crucified in a discordant or provocative manner. Far too often the Cross is presented for our adoration, not so much as a sublime end to be attained by our transcending ourselves, but as a symbol of sadness, of limitation and repression.

This way of preaching the Passion is, in many cases, merely the result of the clumsy use of pious vocabulary in which the most solemn words (sacrifice, immolation, expiation), emptied of their meaning by routine, are used, quite unconsciously, in a light and frivolous way. They become formulas to be juggled with. But this manner of speech ends by conveying the impression that the kingdom of God can only be established in mourning, and by thwarting and going against the current of man's aspirations and engeries. In spite of the verbal fidelity displayed by the use of this kind of language, a picture is presented that is utterly unchristian. What we said just now about the necessary combination of attachment and detachment allows of giving christian asceticism a much richer and a far more complete meaning.

In its highest and most general sense, the doctrine of the Cross is that to which all men adhere who believe that the vast movement and agitation of human life opens on to a road which leads somewhere, and that that road *climbs upward*. Life has a term: therefore it imposes a particular direction, orientated, in fact, towards the highest possible spiritualisation by means of the greatest possible effort. To admit that group of fundamental principles is already to range oneself among the disciples—distant, perhaps, and implicit, but nevertheless real—of Christ crucified. Once that first choice has been made, the first distinction has been drawn between the brave who will succeed and the pleasure-seekers who will fail, between the elect and the condemned.

This rather vague attitude is clarified and carried further by Christianity. Above all, by revealing an original fall, Christianity provides our intelligence with a reason for the disconcerting excess of sin and suffering at certain points. Next, in order to win our love and secure our faith, it unveils to our eyes and hearts the moving and unfathomable reality of the historical Christ in whom the exemplary life of an individual man conceals this mysterious drama: the Master of the world, leading, like an element of the world, not only an elemental life, but (in addition to this and because of it) leading the total life of the universe, which he has shouldered and assimilated by experiencing it himself. And finally by the crucifixion and death of this adored being, Christianity signifies to our thirst for happiness that the term of creation is not to be sought in the temporal zones of our visible world, but that the effort required of our fidelity must be consummated *beyond a total transformation* of ourselves and of everything surrounding us.

Thus the perspectives of renunciation implied in the exercise of life itself are gradually expanded. Ultimately we find ourselves thoroughly uprooted, as the Gospel desires, from everything perceptible on earth. But the process of uprooting ourselves has happened little by little and according to a rhythm which has neither alarmed nor wounded the respect we owe to the admirable beauties of the human effort.

It is perfectly true that the Cross means going beyond the frontiers of the sensible world and even, in a sense, breaking with it. The final stages of the ascent to which it calls us compel us to cross a threshold, a critical point, where we lose touch with the zone of the realities of the senses. That final "excess," glimpsed and accepted from the first steps, inevitably puts everything we do in a special light and gives it a particular significance. That is exactly where the folly of Christianity lies in the eyes of the "wise" who are not prepared to stake the good which they now hold in their hands on a total "beyond." But that agonising flight from the experimental zones—which is what the Cross means—is only (as should be strongly emphasised) the sublime aspect of a law common to *all* life. Towards the peaks, shrouded in mist from our human eyes, whither the Cross beckons us, we rise by a path which is the way of universal progress. The royal road of the Cross is no more nor less than the road of human endeavour supernaturally righted and prolonged. Once we have fully grasped the meaning of the Cross, we are no longer in danger of finding life sad and ugly. We shall simply have become more attentive to its barely comprehensible solemnity.

To sum up, Jesus on the Cross is both the symbol and the reality of the immense labour of the centuries which has, little by little, raised up the created spirit and brought it back to the depths of the divine *milieu*. He represents (and in a true sense, he is) creation, as, upheld by God, it reascends the slopes of being, sometimes clinging to things for support, sometimes tearing itself from them in order to pass beyond them, and always compensating, by physical suffering, for the setbacks caused by its moral downfalls.

The Cross is therefore not inhuman but superhuman. We can now understand that from the very first, from the very origins of mankind as we know it, the Cross was placed on the crest of the road which leads to the highest peaks of creation. But, in the growing light of Revelation, its arms, which at first were bare, show themselves to have put on Christ: *Crux inuncta*. At first sight the bleeding body may seem funereal to us. Is it not from the night that it shines forth? But if we go nearer we shall recognise the flaming Seraph of Alvernus whose passion and compassion are *incendium mentis*. The Christian is not asked to swoon in the shadow, but to climb in the light, of the Cross.

FRENCH EDITOR'S NOTE

In pages that were not, like *Le Milieu Divin,* intended for "the waverers both inside and outside," Père Teilhard, in the course of a meditation, freely expressed the capital importance which he attached to the priestly and religious vocation, to the evangelical counsels, and to the redemptive power of death. The following short extracts will give an idea of the substance of his convictions:

Every priest, because he is a priest, has dedicated his life to the work of universal salvation. If he is conscious of the dignity of his office, he should no longer live for himself but for the world, following the example of him whom he is anointed to represent.

To the full extent of my power, *because I am a priest*, I wish from now on to be the first to become conscious of all that the world loves, pursues and suffers; I want to be the first to seek, to sympathise and to suffer; the first to open myself out and sacrifice myself—to become more widely human and more nobly of the earth than any of the world's servants. . . .

At the same time, by the practice of the counsels and through renunciation, I want to recover all that there may be of heavenly fire in the threefold concupiscence—I want to sanctify, through chastity, poverty and obedience, the power invested in love, in gold and in independence.

Was there ever a humanity, O Lord, more like, in its blood, an immolated victim; more adapted, by its inward unrest, to creative transformations; more rich, with its violence, in sanctifiable energy; more close, in its anguish, to supreme communion? . . .

O priests! Never have you been priests in so full a sense as now, merged and submerged as you are in the pains and the blood of a generation—never so active, never so directly on the path of your vocation. . . .

I feel so weak, Lord, that I hardly dare ask you to let me participate in that beatitude. But I perceive it clearly enough, and I proclaim it:

Happy are those of us who, in these decisive days of the Creation and the Redemption, are chosen for this supreme act, the logical crowning of their priesthood: communion unto death with Christ. . . . (From *Le Prêtre*)

3. THE SPIRITUAL POWER OF MATTER

The same beam of light which christian spirituality, rightly and fully understood, directs upon the Cross to humanise it (without veiling it) is reflected on matter so as to spiritualise it.

In their struggle towards the mystical life, men have often succumbed to the illusion of crudely contrasting soul and body, spirit and flesh, as good and evil. But despite certain current expressions, this Manichean tendency has never had the Church's approval. And, in order to prepare the way for our final view of the divine *milieu*, perhaps we may be allowed to vindicate and exalt that aspect of it which the Lord came to put on, save and consecrate: *holy matter*.

From the mystical and ascetic point of view adopted in these pages, matter is not exactly any of the abstract entities defined under that name by science and philosophy. It is certainly the same *concrete* reality, for us, as it is for physics and metaphysics, having the same basic attributes of plurality, perceivability and inter-connection. But here we want to embrace that reality as a whole in its widest possible sense: to give it its full abundance as it reacts not only to our scientific or analytical investigations, but to all our practical activities. Matter, as far as we are concerned, is the assemblage of things, energies and creatures which surround us in so far as these are palpable, sensible and "natural" (in the theological sense of the word). Matter is the common, universal, tangible setting, infinitely shifting and varied, in which we live.

How, then, does the thing thus defined present itself to us to be acted upon? Under the enigmatic features of a two-sided power.

On the one hand matter is the burden, the fetters, the pain, the sin and the threat to our lives. It weighs us down, suffers, wounds, tempts and grows old. Matter makes us heavy, paralysed, vulnerable, guilty. Who will deliver us from this body of death?

But at the same time matter is physical exuberance, ennobling contact, virile effort and the joy of growth. It attracts, renews, unites and flowers. By matter we are nourished, lifted up, linked to everything else, invaded by life. To be deprived of it is intolerable. *Non exui volumus sed superindui* (2 Cor. v, 4). Who will give us an immortal body?

Asceticism deliberately looks no further than the first aspect, the one which is turned towards death; and it recoils, exclaiming "Flee!" *But what would our spirits be, O God, if they did not have the bread of earthly things to nourish them, the wine of created beauties to intoxicate them, and the conflicts of human life to fortify them? What feeble powers and bloodless hearts your creatures would bring you if they were to succeed in cutting themselves off prematurely from the providential setting in which you have placed them! Teach us, Lord, how to contemplate the sphinx without succumbing to its spell; how to grasp the hidden mystery in the womb of death, not by a refinement of human doctrine, but in the simple concrete act by which you plunged yourself into matter in order to redeem it. By the virtue of your suffering incarnation disclose to us, and then teach us to harness jealousy for you, the spiritual power of matter.*

Let us take a comparison as our starting point. Imagine a deep-sea diver trying to get back from the seabed to the clear light of day. Or imagine a traveller on a fog-bound mountain-side climbing upward towards the summit bathed in light. For each of these men space is divided into two zones marked with opposing properties: the one behind and beneath appears ever darker, while the one in front and above becomes ever lighter. Both diver and climber can succeed in making their way towards the second zone only if they use everything around and about them as points of leverage. Moreover, in the course of their task, the light above them grows brighter with each advance made; and at the same time the area which has been traversed, as it is traversed, ceases to hold the light and is engulfed in darkness. Let us remember these stages, for they express symbolically all the elements we need in order to understand how we should touch and handle matter with a proper sense of reverence.

Above all matter is not just the weight that drags us down, the mire that sucks us in, the bramble that bars our way. In itself, and before we find ourselves where we are, and before we choose, it is simply the slope on which we can go up just as well as go down, the medium that can uphold or give way, the wind that can overthrow or lift up. Of its nature, and as a result of original sin, it is true that it represents a perpetual impulse towards failure. But by nature too, and as a result of the Incarnation, it contains the spur or the allurement to be our accomplice towards heightened being, and this counter-balances and even dominates the *fomes peccati*. The full truth of our situation is that, here below, and by virtue of our immersion in the universe, we are each one of us placed within its layers or on its slopes, at a specific point defined by the present moment in the history of the world, the place of our birth, and our individual vocation. And *from that starting point,* variously situated at different levels, the task assigned to us is to climb towards the light, passing through, so as to attain God, *a given series of created things* which are not exactly obstacles but rather foot-holds, intermediaries to be made use of, nourishment to be taken, sap to be purified and elements to be associated with us and borne along with us.

That being so, and still as a result of our initial position among things, and also as a result of each position we subsequently occupy in matter, matter falls into two distinct zones, differentiated according to our effort: the zone already left behind or arrived at, to which we should not return, or at which we should not pause, lest we fall back—this is the zone of matter *in the material and carnal sense;* and the zone offered to our renewed efforts towards progress, search, conquest and "divinisation," the zone of matter *taken in the spiritual sense;* and the frontier between these two zones is essentially relative and shifting. That which is good, sanctifying and spiritual for my brother below or beside me on the mountainside, can be material, misleading or bad for me. What I rightly allowed myself yesterday, I must perhaps deny myself today. And conversely, actions which would have been a grave betrayal in a St. Aloysius Gonzaga or a St. Anthony, may well be models for me if I am to follow in the footsteps of these saints. In other words, the soul can only rejoin God after having traversed *a specific path* through matter—which path can be seen as the distance which separates, but it can also be seen as the road which links. Without certain possessions and certain victories, no man exists as God wishes him to be. Each one of us has his Jacob's ladder, whose rungs are formed of a series of objects. Thus it is not our business to withdraw from the world before our time; rather let us learn to orientate our being in the flux of things; then, instead of the force of gravity which drags us down to the abyss of self-indulgence and selfishness, we shall feel a salutary "component" emerge from created things which, by a process we have already described, will enlarge our horizons, will snatch us away from our pettinesses and impel us imperiously towards a widening of our vision, towards the renunciation of cherished pleasure, towards the desire for ever more spiritual beauty. Matter, which at first seemed to counsel us towards the maximum pleasure and the minimum effort, emerges as the principle of minimum pleasure and maximum effort.

In this case, too, the law which applies to the individual would seem to be a small-scale version of the law which applies to the whole. It would surely not be far wrong to suggest that, in its universality, the world too has a prescribed path to follow before attaining its consummation. There can really be no doubt of it. If the material totality of the world includes energies which cannot be made use of, and if, more unfortunately, it contains perverted energies and elements which are slowly separated from it, it is still more certain that it contains *a certain quantity of spiritual power* of which the progressive sublimation, *in Christo Jesu,* is, for the Creator, the fundamental operation taking place. At the present time this power is still diffused almost everywhere: nothing, however insignificant or crude it may appear, is without some trace of it. And the task of the body of Christ, living in his faithful, is patiently to sort out those heavenly forces—to extract, without letting any of it be lost, that chosen substance. Little by little, we may rest assured, the work is being done. Thanks to the multitude of individuals and vocations, the Spirit of God insinuates itself everywhere and is everywhere at work. It is the great tree we spoke of a moment ago, whose sunlit branches refine and turn to flowers the sap extracted by the humblest of its roots. As the work progresses, certain zones, no doubt, become worked out. Within each individual life, as we have noted, the frontier between spiritual matter and carnal matter is constantly moving upward. And in the same way, in proportion as humanity is christianised, it feels less and less need for certain earthly nourishment. Contemplation and chastity should thus tend, quite legitimately, to gain mastery over anxious work and direct possession. This is the *general "drift" of matter* towards spirit. This movement must have its term: one day the whole divinisable substance of matter will have passed into the souls of men; all the chosen dynamisms will have been recovered: and then our world will be ready for the Parousia.

Who can fail to perceive the great symbolic gesture of baptism in this general history of matter? Christ immerses himself in the waters of Jordan, symbol of the forces of the earth. These he sanctifies. And as he emerges, in the words of St. Gregory of Nyssa, with the water which runs off his body he elevates the whole world.

Immersion and emergence; participation in things and sublimation; possession and renunciation; crossing through and being borne onwards—that is the twofold yet single movement which answers the challenge of matter in order to save it.*

Matter, you in whom I find both seduction and strength, you in whom I find blandishment and virility, you who can

---

*The sensual mysticisms and certain neo-Pelagianisms (such as Americanism), by paying too much attention to the first of these phases, have fallen into the error of seeking divine love and the divine kingdom *on the same level* as human affections and human progress. Conversely, by concentrating too much on the second phase, some exaggerated forms of Christianity conceive perfection as built upon the destruction of "nature." The true christian supernatural, frequently defined by the Church, neither leaves the creature where he is, on his own plane, nor suppresses him: it "sur-animates" him. It must surely be obvious that, however transcendent and creative they may be, God's love and ardour could only fall upon the *human* heart, that is to say upon an object prepared (from near or from afar) by means of all the nourishments of the earth. It is astonishing that so few minds should succeed, in this as in other cases, in grasping the notion of transformation. Sometimes the thing transformed seems to them to be the old thing unchanged; at other times they see in it only the entirely new. In the first case it is the spirit that eludes them; in the second case, it is the matter. Though not so crude as the first excess, the second is shown by experience to be no less destructive of the equilibrium of mankind.

enrich and destroy, I surrender myself to your mighty layers, with faith in the heavenly influences which have sweetened and purified your waters. The virtue of Christ has passed into you. *Let your attractions lead me forward, let your sap be the food that nourishes me; let your resistance give me toughness; let your robberies and inroads give me freedom. And finally, let your whole being lead me towards Godhead.*

PART THREE

The Divine Milieu

Nemo sibi vivit, aut sibi moritur ... Sive vivimus, sive morimur, Christi sumus.
No man lives or dies to himself. But whether through our life or through our death we belong to Christ.

*The first two parts of this essay are simply an analysis and veri-*fication of the above words of St. Paul. We have considered, in turn, the sphere of activity, development and life, and the sphere of passivity, diminishment and death in our lives. All around us, to right and left, in front and behind, above and below, we have only had to go a little beyond the frontier of sensible appearances in order to see the divine welling up and showing through. But it is not only close to us, in front of us, that the divine presence has revealed itself. It has sprung up so universally, and we find ourselves so surrounded and transfixed by it, that there is no room left to fall down and adore it, even within ourselves.

By means of all created things, without exception, the divine assails us, penetrates us and moulds us. We imagined it as distant and inaccessible, whereas in fact we live steeped in its burning layers. *In eo vivimus.* As Jacob said, awakening from his dream, the world, this palpable world, which we were wont to treat with the boredom and disrespect with which we habitually regard places with no sacred association for us, is in truth a holy place, and we did not know it. *Venite, adoremus.*

Let us withdraw to the higher and more spiritual ether which bathes us in living light. And let us take joy in making an inventory of its attributes and recognising their nature, before examining in a general way the means by which we can open ourselves evermore to its penetration.

1. THE ATTRIBUTES OF THE DIVINE *MILIEU*

The essential marvel of the divine *milieu* is the ease with which it assembles and harmonises within itself qualities which appear to us to be contradictory.

As vast as the world and much more formidable than the most immense energies of the universe, it nevertheless possesses in a supreme degree that precise concentrated particularity which makes up so much of the warm charm of human persons.

Vast and innumerable as the dazzling surge of creatures that are sustained and sur-animated by its ocean, it nevertheless retains the concrete transcendence that allows it to bring back the elements of the world, without the least confusion, within its triumphant and personal unity.

Incomparably near and perceptible—for it presses in upon us through all the forces of the universe—it nevertheless eludes our grasp so constantly that we can never seize it here below except by raising ourselves, uplifted on its waves, to the extreme limit of our effort: present in, and drawing at the inaccessible depth of, each creature, it withdraws always further, bearing us along with it towards the common centre of all consummation.*

---

*I attain God in those whom I love to the same degree in which we, myself and they, become more and more spiritual. In the same way, I grasp

Through it, the touch of matter is a purification, and chastity flowers as the transfiguration of love.

In it, development culminates in renunciation; attachment to things yet separates us from everything disintegrating within them. Death becomes a resurrection.

Now, if we try to discover the source of so many astonishingly coupled perfections, we shall find they all spring from the same "fontal" property which we can express thus: God reveals himself everywhere, beneath our groping efforts, *as a universal milieu,* only because he is *the ultimate point* upon which all realities converge. Each element of the world, whatever it may be, only subsists, *hic et nunc,* in the manner of a cone whose generatrices meet in God who draws them together—(meeting at the term of their individual perfection and at the term of the general perfection of the world which contains them). It follows that all created things, every one of them, cannot be looked at, in their nature and action, without the same reality being found in their innermost being—like sunlight in the fragments of a broken mirror—one beneath its multiplicity, unattainable beneath its proximity, and spiritual beneath its materiality. No object can influence us by its essence without our being touched by the radiance of the focus of the universe. Our minds are incapable of grasping a reality, our hearts and hands of seizing the essentially desirable in it, without our being compelled *by the very structure of things* to go back to the first source of its perfections. This focus, this source, is thus everywhere. It is *precisely because* he is at once so deep and yet so akin to an extensionless point that God is infinitely near, and dispersed everywhere. It is *precisely because* he is the centre that he fills

---

him in the Beautiful and the Good in proportion as I pursue these further and further with progressively purified faculties.

the whole sphere. The omnipresence of the divine is simply the effect of its extreme spirituality and is the exact contrary of the fallacious ubiquity which matter seems to derive from its extreme dissociation and dispersal. In the light of this discovery, we may resume our march through the inexhaustible wonders which the divine *milieu* has in store for us.

However vast the divine *milieu* may be, it is in reality a *centre*. It therefore has the properties of a centre, and above all the absolute and final power to unite (and consequently to complete) all beings within its breast. In the divine *milieu* all the elements of the universe *touch each other* by that which is most inward and ultimate in them. There they concentrate, little by little, all that is purest and most attractive in them without loss and without danger of subsequent corruption. There they shed, in their meeting, the mutual externality and the incoherences which form the basic pain of human relationships. Let those seek refuge there who are saddened by the separations, the meannesses and the wastefulnesses of the world. In the external spheres of the world, man is always torn by the separations which set distance between bodies, which set the impossibility of mutual understanding between souls, which set death between lives. Moreover at every minute he must lament that he cannot pursue and embrace everything within the compass of a few years. Finally, and not without reason, he is incessantly distressed by the crazy indifference and the heart-breaking dumbness of a natural environment in which the greater part of individual endeavour seems wasted or lost, where the blow and the cry seem stifled on the spot, without awakening any echo.

All that desolation is only on the surface.

But let us leave the surface, and, without leaving the world, plunge into God. There, and from there, in him and through him, we shall hold all things and have command of all things. There we shall one day rediscover the essence and brilliance of all the flowers and lights which we were forced to abandon so as to be faithful to life. The beings we despaired of reaching and influencing are all there, all reunited by the most vulnerable, receptive and enriching point in their substance. In this place the least of our desires and efforts is harvested and tended and can at any moment cause the marrow of the universe to vibrate.

Let us establish ourselves in the divine *milieu*. There we shall find ourselves where the soul is most deep and where matter is most dense. There we shall discover, where all its beauties flow together, the ultra-vital, the ultra-sensitive, the ultra-active point of the universe. And, at the same time, we shall feel the *plenitude* of our powers of action and adoration effortlessly ordered within our deepest selves.

But the fact that all the external springs of the world should be co-ordinated and harmonised at that privileged point is not the only marvel. By a complementary marvel, the man who abandons himself to the divine *milieu* feels his inward powers clearly directed and vastly expanded by it with a sureness which enables him to avoid, like child's play, the reefs on which mystical ardour has so often foundered.

In the first place, the sojourner in the divine *milieu* is not a pantheist. At first sight, perhaps, the depths of the divine which St. Paul reveals to us may seem to resemble the fascinating domains unfolded before our eyes by monistic philosophies or religions. In fact they are very different, far more reassuring to our minds, far more comforting to our hearts. Pantheism seduces us by its vistas of perfect universal union. But ultimately, if it were true, it would give us only fusion and unconsciousness; for, at the end of the evolution it claims to reveal, the elements of the world vanish in the God they create or by which they are absorbed. Our God, on

the contrary, pushes to its furthest possible limit the differentiation among the creatures he concentrates within himself.
At the peak of their adherence to him, the elect also discover
in him the consummation of their individual fulfilment.
Christianity alone therefore saves, with the rights of thought,
the essential aspiration of all mysticism: *to be united* (that
is, to become the other) *while remaining oneself.* More attractive than any world-Gods, whose eternal seduction it
embraces, transcends and purifies—*in omnibus omnia Deus*
(*En pasi panta Theos*)—our divine *milieu* is at the antipodes
of false pantheism. The Christian can plunge himself into it
whole-heartedly without the risk of finding himself one day
a monist.

Nor is there any reason to fear that in abandoning himself to those deep waters, he will lose his foothold in revelation and in life, and become either unrealistic in the object of
his worship or else chimerical in the substance of his work.
The Christian lost within the divine layers will not find his
mind subject to the forbidden distortions that go to make
the "modernist" or the "illuminati."

To the Christian's sensitised vision, it is true, the Creator
and, more specifically, the Redeemer (as we shall see) have
steeped themselves in all things and penetrated all things to
such a degree that, as Blessed Angela of Foligno said, "the
world is full of God." But this augmentation is only valuable
in his eyes in so far as the light, in which everything seems to
him bathed, radiates from *a historical centre* and is transmit-
ted along *a traditional and solidly defined axis.* The immense
enchantment of the divine *milieu* owes all its value in the
long run to the human-divine contact which was revealed at
the Epiphany of Jesus. If you suppress the historical reality
of Christ, the divine omnipresence which intoxicates us
becomes, like all the other dreams of metaphysics, uncertain,

vague, conventional—lacking the decisive experimental veri-
fication by which to impose itself on our minds, and without
the moral authority to assimilate our lives into it. Thenceforward, however dazzling the expansions which we shall try in
a moment to discern in the resurrected Christ, their beauty
and their stuff of reality will always remain inseparable from
the tangible and verifiable truth of the Gospel event. The
mystical Christ, the universal Christ of St. Paul, has neither
meaning nor value in our eyes except as an expansion of the
Christ who was born of Mary and who died on the cross.
The former essentially draws his fundamental quality of
undeniability and concreteness from the latter. However far
we may be drawn into the divine spaces opened up to us by
christian mysticism, we never depart from the Jesus of the
Gospels. On the contrary, we feel a growing need to enfold
ourselves ever more firmly within his human truth. We are
not, therefore, modernist in the condemned sense of the
word. Nor shall we end up among the visionaries and the
"illuminati."

The real error of the visionaries is to *confuse* the different
planes of the world, and consequently to mix up their activities. In the view of the visionary, the divine presence illumi-
nates not only the heart of things, but tends to invade their
surface and hence to do away with their exacting but salutary
reality. The gradual maturing of immediate causes, the deter-
minate systems of material order in their complex inter-
relationships, the infinite susceptibilities of the universal
order, no longer count. Through this veil without seam and
these delicate threads, divine action is imagined as appearing
naked and without order. And then the falsely miraculous
comes to disconcert and obstruct the human effort.

As we have already abundantly shown, the effect produced upon human activity by the true transformation of the

world in Jesus Christ, is utterly different. At the heart of the divine *milieu,* as the Church reveals it, things are transfigured, but from within. They bathe inwardly in light, but, in this incandescence, they retain—this is not strong enough, they exalt—all that is most specific in their attributes. *We can only lose ourselves in God by prolonging the most individual characteristics of beings far beyond themselves:* that is the fundamental rule by which we can always distinguish the true mystic from his counterfeits. The heart of God is boundless, *multae mansiones.* And yet in all that immensity there is only one possible place for each one of us at any given moment, the one we are led to by unflagging fidelity to the natural and supernatural duties of life. At this point, which we can reach at the right moment only if we exert the maximum effort on every plane, God will reveal himself in all his plenitude. Except at this point, the divine *milieu,* although it may still enfold us, exists only incompletely, or not at all, *for us.* Thus its great waters do not call us to defeat but to perpetual struggle to breast their floods. Their energy awaits and provokes our energy. Just as on certain days the sea lights up only as the ship's prow or the swimmer cleaves its surface, so the world is lit up with God only when reacting to our impetus. When God desires ultimately to subject and unite the Christian to him, either by ecstasy or by death, it is as though he bears him away stiffened by love and by obedience in the full extent of his effort.

It might thenceforward look as though the believer in the divine *milieu* were falling back into the errors of a pagan naturalism in reaction against the excesses of quietism and illuminism. With his faith in the heavenly value of human endeavour, by his expectation of a new awakening of the faculties of adoration dormant in the world, by his respect for the spiritual powers still latent in matter, the Christian may often bear a striking resemblance to the worshippers of the earth.

But here again, as in the case of pantheism, the resemblance is only external and *such as is so often found in opposite things.*

The pagan loves the earth in order to enjoy it and confine himself within it; the Christian in order to make it purer and draw from it the strength to escape from it.

The pagan seeks to espouse sensible things so as to extract delight from them; *he adheres to the world.* The Christian multiplies his contacts with the world only so as to harness, or submit to, the energies which he will take back, or which will take him, to heaven. *He pre-adheres to God.*

The pagan holds that man divinises himself by closing in upon himself; the final act of human evolution is when the individual, or the totality, constitutes itself within itself. The Christian sees his divinisation only in the assimilation by an "Other" of his achievement: the culmination of life, in his eyes, is death in union.

To the pagan, universal reality exists only in so far as it is projected on to the plane of the perceptible: it is immediate and multiple. The Christian makes use of exactly the same elements: but he prolongs them along their common axis, which links them to God: and, by the same token, the universe is thus unified for him, although it is only attainable at the final centre of its consummation.

To sum up, one may say that, in relation to all the main historical forms assumed by the human religious spirit, christian mysticism extracts *all* that is sweetest and strongest circulating in all the human mysticisms, though without absorbing their evil or suspect elements. It shows an astonishing equilibrium between the active and the passive, between possession of the world and its renunciation, between a taste for things and an indifference to them. But there is really no reason why we should be astonished by this shifting harmony, for is it not the natural and spontaneous reaction of the soul to the stimulus of a *milieu* which is exactly, by nature and grace, the one in which that soul is made to live and develop itself? Just as, at the centre of the divine *milieu*, all the sounds of created being are fused, without being confused, in a single note which dominates and sustains them (that seraphic note, no doubt, which bewitched St. Francis), so all the powers of the soul begin to resound in response to its call; and these multiple tones, in their turn, compose themselves into a single, ineffably simple vibration in which all the spiritual nuances—of love and of the intellect, of zeal and of tranquillity, of fullness and of ecstasy, of passion and of indifference, of assimilation and of surrender, of rest and of motion—are born and pass and shine forth, according to the times and the circumstances, like the countless possibilities of an inward attitude, inexpressible and unique.

And if any words could translate that permanent and lucid intoxication better than others, perhaps they would be "passionate indifference."

To have access to the divine *milieu* is to have found the one thing needful: *him who burns* by setting fire to everything that we would love badly or not enough; *him who calms* by eclipsing with his blaze everything that we would love too much; *him who consoles* by gathering up everything that has been snatched from our love or has never been given to it. To reach those priceless layers is to experience, with equal truth, that one has need of everything, and that one has need of nothing. Everything is needed because the world will never be large enough to provide our taste for action with the means of grasping God, or our thirst for undergoing with the possibility of being invaded by him. And yet nothing is needed; for as the only reality which can satisfy us lies beyond the transparencies in which it is mirrored, everything that fades away and dies between us will only serve to give reality back to us with greater purity. Everything means both everything and nothing to me; everything is God to me and everything is dust to me: that is what man can say with equal truth, in accord with how the divine ray falls.

"Which is the greater blessing," someone once asked, "to have the sublime unity of God to centre and save the universe? or to have the concrete immensity of the universe by which to undergo and touch God?"

We shall not seek to escape this joyful uncertainty. But now that we are familiar with the attributes of the divine *milieu*, we shall turn our attention to the Thing itself which appeared to us in the depth of each being, like a radiant countenance, like a fascinating abyss. We can now say "Lord, who art thou?"

2. THE NATURE OF THE DIVINE *MILIEU*
THE UNIVERSAL CHRIST
AND THE GREAT COMMUNION

We can say as a first approximation that the *milieu* whose rich and mobile homogeneity has revealed itself all around us as a condition and a consequence of the most christian attitudes (such as right intention and resignation) is formed by the divine omnipresence. The immensity of God is the essential attribute which allows us to seize him everywhere, within us and around us.

This answer begins to satisfy our minds in that it circumscribes the problem. However, it does not give to the power *in quo vivimus et sumus* the sharp lines with which we should wish to trace the features of the one thing needful. Under what form, proper to our creation and adapted to our universe, does the divine immensity manifest itself to, and become relevant to, mankind? We feel it charged with that sanctifying grace which the Catholic faith causes to circulate everywhere as the true sap of the world; which, in its attributes, is very like that charity *(manete in dilectione mea)* which will one day, the Scriptures tell us, be the only stable principle of natures and powers; which, too, is fundamentally similar to the wonderful and substantial divine will, whose marrow is everywhere present and constitutes the true food of our lives, *omne delectamentum in se habentem.* What is, when all is said and done, the concrete link which binds all these universal entities together and confers on them a final power of gaining hold of us?

The essence of Christianity consists in asking oneself that question, and in answering: "The Word incarnate, our Lord Jesus Christ."

Let us examine step by step how we can validate to ourselves this prodigious identification of the Son of Man and the divine *milieu.*

A first step, unquestionably, is to see the divine omnipresence in which we find ourselves plunged as *an omnipresence of action.* God enfolds us and penetrates us by creating and preserving us.

Now let us go a little further. Under what form, and with what end in view, has the Creator given us, and still preserves in us, the gift of participated being? Under the form of an essential aspiration towards him—and with a view to the unhoped-for cleaving which is to make us one and the same complex thing with him. The action by which God maintains us in the field of his presence is *a unitive transformation*.

Let us go further still. What is the supreme and complex reality for which the divine operation moulds us? It is revealed to us by St. Paul and St. John. It is the quantitative repletion and the qualitative consummation of all things: it is the mysterious Pleroma, in which the substantial *one* and the created *many* fuse without confusion in a *whole* which, without adding anything essential to God, will nevertheless be a sort of triumph and generalisation of being.

At last we are nearing our goal. What is the active centre, the living link, the organising soul of the Pleroma? St. Paul, again, proclaims it with all his resounding voice: it is he in whom everything is reunited, and in whom all things are consummated—through whom the whole created edifice receives its consistency—Christ dead and risen *qui replet omnia, in quo omnia constant*.

And now let us link the first and last terms of this long series of identities. We shall then see with a wave of joy that *the divine omnipresence* translates itself within our universe by the network of the organising forces of the total Christ. God exerts pressure, in us and upon us—through the intermediary of all the powers of heaven, earth and hell—only in the act of forming and consummating Christ who saves and sur-animates the world. And since, in the course of this operation, Christ himself does not act as a dead or passive point of convergence, but as a centre of radiation for the energies which lead the universe back to God through his humanity, the layers of divine action finally come to us impregnated with his organic energies.

The divine *milieu* henceforward assumes for us the savour and the specific features which we desire. In it we recognise an omnipresence which acts upon us by assimilating us in it, *in unitate corporis Christi.* As a consequence of the Incarnation, the divine immensity has transformed itself for us into *the omnipresence of christification.* All the good that I can do *opus et operatio* is physically gathered in, by something of itself, into the reality of the consummated Christ. Everything I endure, with faith and love, by way of diminishment or death, makes me a little more closely an integral part of his mystical body. Quite specifically it is *Christ whom we make or whom we undergo in all things.* Not only *diligentibus omnia convertuntur in bonum* but, more clearly still, *convertuntur in Deum* and, quite explicitly, *convertuntur in Christum.*

In spite of the strength of St. Paul's expressions (formulated, it should be remembered, for the *ordinary run* of the first Christians) some readers may feel that we have been led to strain, in too realist a direction, the meaning of "mystical body"—or at least that we have allowed ourselves to seek esoteric perspectives in it. But if we look a little more closely, we shall see that we have simply taken another path in order to rejoin the great highway opened up in the Church by the onrush of the cult of the Holy Eucharist.

When the priest says the words *Hoc est Corpus meum,* his words fall directly on to the bread and directly transform it into the individual reality of Christ. But the great sacramental operation does not cease at that local and momentary event. Even children are taught that, throughout the life of each man and the life of the Church and the history of the world, there is only one Mass and one Communion. Christ died once in agony. Peter and Paul receive communion on such and such a day at a particular hour. But these different acts are only the diversely central points in which the continuity of a unique act is split up and fixed, in space and time, for our experience. In fact, from the beginning of the Messianic preparation, up till the Parousia, passing through the historic manifestation of Jesus and the phases of growth of his Church, a single event has been developing in the world: the Incarnation, realised, in each individual, through the Eucharist.

All the communions of a life-time are one communion.

All the communions of all men now living are one communion.

All the communions of all men, present, past and future, are one communion.

Have we ever sufficiently considered the physical immensity of man, and his extraordinary relations with the universe, in order to realise in our minds the formidable implications of this elementary truth?

Let us conjure up in our minds, as best we can, the vast multitudes of men in every epoch and in every land. According to the catechism we believe that this fearful anonymous throng is, by right, subject to the physical and overmastering contact of him whose appanage it is to be able *omnia sibi subicere* (by right, and to a certain extent in fact; for who can tell where the diffusion of Christ, with the influence of grace, stops, as it spreads outward from the faithful at the heart of the human family?). Yes, the human layer of the earth is wholly and continuously under the organising influx of the incarnate Christ. This we all believe, as one of the most certain points of our faith.

Now how does the human world itself appear within the structure of the universe? We have already spoken of this (pp. 21 ff.), and the more one thinks of it the more one is struck by the obviousness and importance of the following conclusion: it appears as a zone of continuous spiritual transformation, where all inferior realities and forces without exception are sublimated into sensations, feelings, ideas and the powers of knowledge and love. Around the earth, the centre of our field of vision, the souls of men form, in some manner, the incandescent surface of matter plunged in God. From the dynamic and biological point of view it is quite as impossible to draw a line below it, as to draw a line between a plant and the environment that sustains it. If, then, the Eucharist is a sovereign influence upon our human natures, then its energy necessarily extends, owing to the effects of continuity, into the less luminous regions that sustain us; *descendit ad inferos,* one might say. At every moment the eucharistic Christ controls—from the point of view of the organisation of the Pleroma (which is the only true point of view from which the world can be understood)—the whole movement of the universe: the Christ *per quem omnia, Domine, semper creas, vivificas et praestas nobis.*

The control of which we are speaking is, at the minimum, a final refinement, a final purification, a final harnessing, of all the elements which can be used in the construction of the New Earth. But how can we avoid going further and believing that the sacramental action of Christ, *precisely because it sanctifies matter,* extends its influence beyond the pure supernatural, over all that makes up the internal and external ambience of the faithful, that is to say that it sets its mark in everything which we call "our providence"?

If this is the case, then we find ourselves (by simply having followed the "extensions" of the Eucharist) plunged once again precisely into our divine *milieu.* Christ—for whom and in whom we are formed, each with his own individuality and his own vocation—Christ reveals himself in each reality around us, and shines like an ultimate determinant, like a centre, one might almost say like a universal element. As our humanity assimilates the material world, and as the Host assimilates our humanity, the eucharistic transfor-

mation goes beyond and completes the transubstantiation of
the bread on the altar. Step by step it irresistibly invades the
universe. It is the fire that sweeps over the heath; the stroke
that vibrates through the bronze. In a secondary and generalised sense, but in a true sense, the sacramental Species are
formed by the totality of the world, and the duration of the
creation is the time needed for its consecration. *In Christo
vivimus, movemur et sumus.*

*Grant, O God, that when I draw near to the altar to com-
municate, I may henceforth discern the infinite perspectives
hidden beneath the smallness and the nearness of the Host in
which you are concealed. I have already accustomed myself to
seeing, beneath the stillness of that piece of bread, a devouring
power which, in the words of the greatest doctors of your Church,
far from being consumed by me, consumes me. Give me the
strength to rise above the remaining illusions which tend to
make me think of your touch as circumscribed and momentary.*

*I am beginning to understand: under the sacramental
Species it is primarily through the "accidents" of matter that
you touch me, but, as a consequence, it is also through the whole
universe in proportion as this ebbs and flows over me under
your primary influence. In a true sense the arms and the heart
which you open to me are nothing less than all the united powers
of the world which, penetrated and permeated to their depths
by your will, your tastes and your temperament, converge upon
my being to form it, nourish it and bear it along towards the
centre of your fire. In the Host it is* my life *that you are offering
me, O Jesus.*

*What can I do to gather up and answer that universal
and enveloping embrace?* Quomodo comprehendam ut com-
prehensus sim? *To the total offer that is made me, I can only
answer by a total acceptance. I shall therefore* react *to the
eucharistic contact with* the entire effort of my life—*of my life*

of today and of my life of tomorrow, of my personal life and of my life as linked to all other lives. Periodically, the sacred Species may perhaps fade away in me. But each time they will leave me a little more deeply engulfed in the layers of your omnipresence: living and dying, I shall never at any moment cease to move forward in you. Thus the precept implicit in your Church, that we must communicate everywhere and always, is justified with extraordinary force and precision. The Eucharist must invade my life. My life must become, as a result of the sacrament, an unlimited and endless contact with you—that life which seemed, a few moments ago, like a baptism with you in the waters of the world, now reveals itself to me as communion with you through the world. It is the sacrament of life. The sacrament of my life—*of my life received, of my life lived, of my life surrendered. . . .*

Because you ascended into heaven after having descended into hell, you have so filled the universe in every direction, Jesus, that henceforth it is blessedly impossible for us to escape you. Quo ibo a spiritu tuo, et quo a facie tua fugiam. *Now I know that for certain. Neither life, whose advance increases your hold upon me; nor death, which throws me into your hands; nor the good or evil spiritual powers which are your living instruments; nor the energies of matter into which you have plunged; nor the irreversible stream of duration whose rhythm and flow you control without appeal; nor the unfathomable abysses of space which are the measure of your greatness,* neque mors, neque vita, neque angeli, neque principatus, neque potestates, neque virtutes, neque instantia, neque futura, neque fortitudo, neque altitudo, neque profundum, neque ulla creatura (Rom. viii, 38)—*none of these things will be able to separate me from your substantial love, because they are all only the veil,* the "species," (under which you take hold of me in order that I may take hold of you.

Once again, Lord, I ask which is the most precious of these two beatitudes: that all things for me should be a contact with you? or that you should be so "universal" that I can undergo you and grasp you in every creature?

Sometimes people think that they can increase your attraction in my eyes by stressing almost exclusively the charm and goodness of your human life in the past. But truly, O Lord, if I wanted to cherish only a man, then I would surely turn to those whom you have given me in the allurement of their present flowering. Are there not, with our mothers, brothers, friends and sisters, enough irresistibly lovable people around us? Why should we turn to Judaea two thousand years ago? No, what I cry out for, like every being, with my whole life and all my earthly passion, is something very different from an equal to cherish: it is a God to adore.

To adore . . . That means to lose oneself in the unfathomable, to plunge into the inexhaustible, to find peace in the incorruptible, to be absorbed in defined immensity, to offer oneself to the fire and the transparency, to annihilate oneself in proportion as one becomes more deliberately conscious of oneself, and to give of one's deepest to that whose depth has no end. Whom, then, can we adore?

The more man becomes man, the more will he become prey to a need, a need that is always more explicit, more subtle and more magnificent, the need to adore.

Disperse, O Jesus, the clouds with your lightning! Show yourself to us as the Mighty, the Radiant, the Risen! Come to us once again as the Pantocrator who filled the solitude of the cupolas in the ancient basilicas! Nothing less than this Parousia is needed to counter-balance and dominate in our hearts the

*glory of the world that is coming into view. And so that we
should triumph over the world with you, come to us clothed in
the glory of the world.*

3. THE GROWTH OF THE DIVINE *MILIEU*

The kingdom of God is within us. When Christ appears in
the clouds he will simply be manifesting a metamorphosis
that has been slowly accomplished under his influence in the
heart of the mass of mankind. In order to hasten his coming,
let us therefore concentrate upon a better understanding of
the process by which the holy presence is born and grows
within us. In order to foster its progress more intelligently let
us observe the birth and growth of the divine *milieu*, first in
ourselves and then in the world that begins with us.

A. *The coming of the divine* milieu. *The taste
for being and the diaphany of God*

A breeze passes in the night. When did it spring up? Whence
does it come? Whither is it going? No man knows. No one
can compel the spirit, the gaze or the light of God to descend
upon him.

On some given day a man suddenly becomes conscious
that he is alive to a particular perception of the divine spread
everywhere about him. Question him. When did this state
begin for him? He cannot tell. All he knows is that a new
spirit has crossed his life.

"It began with a particular and unique resonance which
swelled each harmony, with a diffused radiance which haloed
each beauty. . . . All the elements of psychological life were
in turn affected; sensations, feelings, thoughts. Day by day

they became more fragrant, more coloured, more intense by means of an indefinable thing—the same thing. Then the vague note, and fragrance, and light began to define themselves. And then, contrary to all expectation and all probability, I began to feel what was ineffably common to all things. The unity communicated itself to me by giving me the gift of grasping it. I had in fact acquired a new sense, *the sense of a new quality* or *of a new dimension*. Deeper still: a transformation had taken place for me *in the very perception of being*. Thenceforward being had become, in some way, tangible and savorous to me; and as it came to dominate all the forms which it assumed, being itself began to draw me and to intoxicate me."

That is what any man might say, more or less explicitly, who has gone any distance in the development of his capacity for self-analysis. Outwardly he could well be a pagan. And should he happen to be a Christian, he would admit that this inward reversal seemed to him to have occurred within the profane and "natural" parts of his soul.

But we must now allow ourselves to be deceived by appearances. We must not let ourselves be disconcerted by the patent errors into which many mystics have fallen in their attempt to place or even to name the universal Smile. As with all power (and the richer, the more so) the sense of the All comes to birth inchoate and troubled. It often happens that, like children opening their eyes for the first time, men do not accurately place the reality which they sense behind things. Their gropings often meet with nothing but a metaphysical phantom or a crude idol. But images and reflections have never proved anything against the reality of objects and of the light. The false trails of pantheism bear witness to our immense need for some revealing word to come from the mouth of him who is. With that reservation, it remains true that, physiologically, the so-called "natural" taste for being is, in each life, the first dawn of the divine illumination—the first tremor perceived of the world animated by the Incarnation. The sense *(which is not necessarily the feeling)* of the omnipresence of God prolongs, sur-creates and supernaturalises the identical physiological energy which, in a mutilated or misdirected form, produces the various styles of pantheism.*

Once we realise that the *divine* milieu *discloses itself to us as a modification of the deep being of things,* it is at once possible to make two important observations touching the manner in which its perception is introduced and preserved within our human horizons.

In the first place, the manifestation of the divine no more modifies the apparent order of things than the eucharistic consecration modifies the sacred Species to our eyes. Since the psychological event consists, at first, solely in the appearance of an *inward tension* or *deep brilliance,* the relations between creatures remains exactly the same. They are merely accentuated in meaning. Like those translucent materials which a light within them can illuminate as a whole, the world appears to the christian mystic bathed in an inward light which intensifies its relief, its structure and its depth. This light is not the superficial glimmer which can be realised in coarse enjoyment. Nor is it the violent flash which destroys objects and blinds our eyes. It is the calm and powerful radi-

---

*In other words and more simply: Just as in the love of God (charity) can be found, quite obviously, the human power to love in its supernatural state—so, in the same way, we believe that at the psychological origin of the "feeling of omnipresence," experienced by the Christian, can be found "the sense of universal Being" which is the source of the majority of human mysticisms. There is a soul which is *naturaliter christiana.* It should be remembered (cf. the Introduction) that these pages contain a psychological description, not a theological explanation, of the states of soul met with.

ance engendered by the synthesis of all the elements of the world in Jesus. The more fulfilled, according to their nature, are the beings in whom it comes to play, the closer and more sensible this radiance appears; and the more sensible it becomes, the more the objects which it bathes become distinct in contour and remote in substance. If we may slightly alter a hallowed expression, we could say that the great mystery of Christianity is not exactly the appearance, but the transparence, of God in the universe. *Yes, Lord, not only the ray that strikes the surface, but the ray that penetrates, not only your Epiphany, Jesus, but your* diaphany.

Nothing is more consistent or more fleeting—more fused with things or at the same time more separable from them—than a ray of light. If the divine *milieu* reveals itself to us as an incandescence of the inward layers of being, who is to guarantee us the persistence of this vision? No-one other than the ray of light itself. The diaphany . . . No power in the world can prevent us from savouring its joys because it happens at a level deeper than any power; and no power in the world—for the same reason—can compel it to appear.

That is the second point, the consideration of which should be used as the basis for all our further reflections on the progress of life in God.

The perception of the divine omnipresence is essentially a seeing, a taste, that is to say a sort of intuition bearing upon certain superior qualities in things. It cannot, therefore, be attained directly by any process of reasoning, nor by any human artifice. It is a gift, like life itself, of which it is undoubtedly the supreme experimental perfection. And so we are brought back again to the centre of ourselves, to the edge of that mysterious source to which we descended (at the beginning of Part Two) and watched it as it welled up. To experience the attraction of God, to be sensible of the beauty, the consistency and the final unity of being, is the highest and at the same time the most complete of our "passivities of growth." God tends, by the logic of his creative effort, to make himself sought and perceived by us: *Posuit homines . . . si forte attrectent eum.* His prevenient grace is therefore always on the alert to excite our first look and our first prayer. But in the end the initiative, the awakening, always come from him, and whatever the further developments of our mystical faculties, no progress is achieved in this domain except as the new response to a new gift. *Nemo venit ad me, nisi Pater traxerit eum.*

We are thus led to posit intense and continual prayer at the origin of our invasion by the divine *milieu*, the prayer which begs for the fundamental gift: Domine, fac ut videam. *Lord, we know and feel that you are everywhere around us; but it seems that there is a veil before our eyes.* Illumina vultum tuum super nos—*let the light of your countenance shine upon us in its universality.* Sit splendor Domini nostri super nos—*may your deep brilliance light up the innermost parts of the massive obscurities in which we move. And, to that end, send us your spirit,* Spiritus principalis, *whose flaming action alone can operate the birth and achievement of the great metamorphosis which sums up all inward perfection and towards which your creation yearns:* Emitte Spiritum tuum, et creabuntur, et RENOVABIS FACIEM TERRAE.

B. *Individual progress in the divine* milieu: *purity, faith and fidelity—the operatives*

Ego operor . . . Pater semper operatur. The delight of the divine *milieu* (heavy with responsibilities) is that it can assume an *ever-increasing* intensity around us. One could say that it is an atmosphere ever more luminous and ever more charged with God. It is in him and in him alone that the reckless vow of all love is realised: to lose oneself in what one loves, to sink oneself in it more and more.

It could be said that three virtues contribute with particular effectiveness towards the limitless concentration of the divine in our lives—purity, faith and fidelity; three virtues which appear to be "static" but which are in fact the three most active and unconfined virtues of all. Let us look at them one after the other and examine their generative function in the divine *milieu*.

i. Purity

Purity, in the wide sense of the word, is not merely abstaining from wrong (that is only a negative aspect of purity), nor even chastity (which is only a remarkable special instance of it). It is the rectitude and the impulse introduced into our lives by the love of God sought in and above everything.

He is spiritually impure who, lingering in pleasure or shut up in selfishness, introduces, within himself and around himself, a principle of slowing-down and division in the unification of the universe in God.

He is pure, on the other hand, who, in accord with his place in the world, seeks to give Christ's desire to consummate all things precedence over his own immediate and momentary advantage.

Still purer and more pure is he who, attracted by God, succeeds in giving that movement and impulse of Christ's an ever greater continuity, intensity and reality—whether his vocation calls him to move always in the material zones of the world (though more and more spiritually), or whether, as is more often the case, he has access to regions where the divine gradually replaces for him all other earthly nourishment.

Thus understood, the purity of beings is measured by the degree of the attraction that draws them towards the divine centre, or, what comes to the same thing, by their proximity to the centre. Christian experience teaches us that it is preserved by recollection, mental prayer, purity of conscience, purity of intention, and the sacraments. Let us be satisfied, here, with extolling its wonderful power of condensing the divine in all around us.

In one of his stories, Robert Hugh Benson tells of a "visionary" coming on a lonely chapel where a nun is praying. He enters. All at once he sees the whole world bound up and moving and organising itself around that out-of-the-way spot, in tune with the intensity and inflection of the desires of that puny, praying figure. The convent chapel had become the axis about which the earth revolved. The contemplative sensitised and animated all things because she believed; and her faith was operative because her very pure soul placed her near to God. This piece of fiction is an admirable parable.

The inward tension of the mind towards God may seem negligible to those who try to calculate the quantity of energy accumulated in the mass of humanity.

And yet, if we could see the "light invisible" as we can see clouds or lightning or the rays of the sun, a pure soul would seem as active in this world, by virtue of its sheer purity, as the snowy summits whose impassable peaks breathe in continually for us the roving powers of the high atmosphere.

If we want the divine *milieu* to grow all around us, then we must jealously guard and nourish all the forces of union, of desire, and of prayer that grace offers us. By the mere fact that our transparency will increase, the divine light, that never ceases to press in upon us, will irrupt the more power-fully.

Have we ever thought of the meaning of the mystery of the Annunciation?

When the time had come when God resolved to realise his incarnation before our eyes, he had first of all to raise up in the world a virtue capable of drawing him as far as our-selves. He needed a mother who would engender him in the human sphere. What did he do? He created the Virgin Mary, that is to say he called forth on earth a purity so great that, within this transparency, he would concentrate himself to the point of appearing as a child.

There, expressed in its strength and reality, is the power of purity to bring the divine to birth among us.

And yet the Church, addressing the Virgin Mother, adds: *Beata quae credidisti*. For it is in faith that purity finds the fulfilment of its fertility.

ii. Faith

Faith, as we understand it here, is not—of course—simply the intellectual adherence to christian dogma. It is taken in a much richer sense to mean belief in God charged with all the trust in his beneficent strength that the knowledge of the divine Being arouses in us. It means the practical conviction that the universe, between the hands of the Creator, still con-tinues to be the clay in which he shapes innumerable possibil-ities according to his will. In a word, it is *evangelical faith*, of which it can be said that no virtue, not even charity, was more strongly urged by the Saviour.

Now, under what guise was this disposition so untiringly revealed to us by the words and deeds of the Master? Above all and beyond all as *an operative power*. But, intimidated by the assertions of an unproven positivism, or "put off" by the mystical excesses of Christian Science, we are sometimes tempted to gloss over the disconcerting promise that the efficacy of prayer is tangible and certain. Yet we cannot ignore it without blushing for Christ. If we do not believe, the waves engulf us, the winds blow, nourishment fails, sickness lays us low or kills us, the divine power is impotent or remote. If, on the other hand, we believe, the waters are welcoming and sweet, the bread is multiplied, our eyes open, the dead rise again, the power of God is, as it were, drawn from him by force and spreads throughout all nature. One must either arbitrarily minimise or explain away the Gospel, or one must admit the reality of these effects not as transient and past, but as perennial and true at this moment. Let us beware of stifling this revelation of a possible vitalisation of the forces of nature in God. Let us, rather, place it resolutely at the centre of our vision of the world—careful, only, that we understand it aright.

When we say that faith is "operative," what do we mean? Is divine action, at the call of faith, going to replace the normal interplay of the causes which surround us? Do we, like the "illuminati," expect God to bring about directly, upon matter or upon our bodies, results that have hitherto been obtained by our own industrious research?

Obviously not. Neither the internal inter-relations of the material or psychical world, nor man's duty to make the greatest possible effort, are in any way undermined, or even relaxed, by the precepts of faith. *Iota unum aut unus apex non praeteribit.* All the natural links of the world remain intact under the transforming action of "operative faith"; but a principle, an inward finality, one might almost say an additional soul, is superimposed upon them. Under the influence of our faith, the universe is capable, without outwardly changing its characteristics, of becoming more supple, more fully animate—of being "sur-animated." That is the "at the most" and the "at the least" of the belief expressly imposed upon us by the Gospel. Sometimes this "sur-animation" expresses itself in miraculous effects—when the transfiguration of causes permits them access to the zone of their "obediential potency." At other times, and this is the more usual case, it is manifested by the integration of unimportant or unfavourable events within a higher plane and within a higher providence.

We have already mentioned and analysed (p. 53) a very typical example of this second form of divinisation of the world by faith (a form no less profound and no less precious than more striking prodigies). In considering the passivities of diminishment we saw how our failures, our death, our faults even, could—through God—be recast into something better and transformed in him. The moment has come to envisage this miracle in its most general sense and from the particular point of view of the act of faith which is, on our part, its providential condition.

In our hands, in the hands of all of us, the world and life *(our world, our life)* are placed like a Host, ready to be charged with the divine influence, that is to say with a real presence of the incarnate Word. The mystery will be accomplished. But on one condition: which is that *we shall believe* that *this* has the will and the power to become for us the action—that is to say the prolongation of the Body of Christ. If we believe, then everything is illuminated and takes shape around us: chance is seen to be order, success assumes an incorruptible plenitude, suffering becomes a visit and a caress of God. But if we hesitate, the rock remains dry, the sky dark, the waters treacherous and shifting. And we may hear the voice of the Master, faced with our bungled lives: "O men of little faith, why have you doubted . . . ?"

Domine, adjuva incredulitatem meam. *Ah, you know it yourself, Lord, through having borne the anguish of it as man: on certain days the world seems a terrifying thing: huge, blind and brutal. It buffets us about, drags us along, and kills us with complete indifference. Heroically, it may truly be said, man has contrived to create a more or less habitable zone of light and warmth in the midst of the great, cold, black waters—a zone where people have eyes to see, hands to help, and hearts to love. But how precarious that habitation is! At any moment the vast and horrible thing may break in through the cracks—the thing which we try hard to forget is always there, separated from us by a flimsy partition: fire, pestilence, storms, earthquakes, or the unleashing of dark moral forces—these callously sweep away in one moment what we had laboriously built up and beautified with all our intelligence and all our love.*

Since my dignity as a man, O God, forbids me to close my eyes to this—like an animal or a child—that I may not succumb to the temptation to curse the universe and him who made it, teach me to adore it by seeing you concealed within it. O Lord, repeat to me the great liberating words, the words which at once reveal and operate: Hoc est Corpus meum. *In truth, the huge and dark thing, the phantom, the storm—if we want it to be so, is you!* Ego sum, nolite timere. *The things in our life which terrify us, the things that threw you yourself into agony in the garden, are, ultimately, only the species or appearance, the matter of one and the same sacrament.*

We have only to believe. And the more threatening and irreducible reality appears, the more firmly and desperately must we believe. Then, little by little, we shall see the universal horror unbend, and then smile upon us, and then take us in its more than human arms.

No, it is not the rigid determinism of matter and of large

numbers, but the subtle combinations of the spirit, that give the universe its consistency. The immense hazard and the immense blindness of the world are only an illusion to him who believes. *Fides, substantia rerum.*

iii. Fidelity

Because we have believed intensely and with a pure heart in the world, the world will open the arms of God to us. It is for us to throw ourselves into these arms so that the divine *milieu* should close around our lives like a circle. That gesture of ours will be one of an active response to our daily task. *Faith consecrates the world. Fidelity communicates with it.*

To give a worthy description of the "advantages" of fidelity, that is to say of the essential and final part which it plays in our taking possession of the divine *milieu*, we should have to go back to what was said in the first two parts of this study. For it is fidelity which releases the inexhaustible resources offered by every "passion" to our desire for communion.

Through fidelity we situate ourselves and maintain ourselves in the hands of God so exactly as to become one with them in their action.

Through fidelity we open ourselves so intimately and continuously to the wishes and good pleasure of God, that his life penetrates and assimilates ours like a fortifying bread. *Hoc est cibus meus, ut faciam voluntatem Patris.*

And finally, through fidelity we find ourselves at every moment situated at the exact point at which the whole bundle of inward and outward forces of the world converge providentially upon us, that is to say at the one point where the divine *milieu* can, at a given moment, be made real for us.

It is fidelity and fidelity alone that enables us to welcome the universal and perpetual overtures of the divine *milieu;*

through fidelity and fidelity alone can we return to God the kiss he is for ever offering us across the world.

What is without price in the "communicating" power of fidelity is that, like the power possessed by faith and purity, it knows no limits to its efficacy.

There is no limit *in respect of the work* done or the diminishment undergone, because we can always sink ourselves deeper into the perfecting of work to be achieved, or into the better utilisation of distressing events. We can always be more industrious, more meticulous, more flexible. . . .

Nor is there any limit *in respect of the intention* which animates our endeavour to act or to accept, because we can always go further in the inward perfecting of our conformity. There can always be greater detachment and greater love.

And there is no limit, indeed there is still less limit, *in respect of the divine object* in the ever-closer espousal of which our being can joyfully wear itself away. This is the moment to abandon all conception of static adherence; it can only be inadequate. And let us remember this: God does not offer himself to our finite beings as a thing all complete and ready to be embraced. For us he is eternal discovery and eternal growth. The more we think we understand him, the more he reveals himself as otherwise. The more we think we hold him, the further he withdraws, drawing us into the depths of himself. The nearer we approach him through all the efforts of nature and grace, the more he increases, in one and the same movement, his attraction over our powers, and the receptivity of our powers to that divine attraction.

Thus the privileged point which was mentioned a short time back—the one point at which the divine *milieu* may be born, for each man, at any moment—is not a fixed point in the universe, but a moving centre which we have to follow, like the Magi their star.

That star leads each man differently, by a different path, in accord with his vocation. But all the paths which it indicates have this in common: that they lead always upward. (We have already said these things more than once, but it is important to group them together for the last time in the same bundle.) In any existence, if it has fidelity, greater desires follow on lesser ones, renunciation gradually gains mastery over pleasure, death consummates life. Finally the general drift throughout creation will have been the same for all. Sometimes through detachment of mind, sometimes through effective detachment, fidelity leads us all, more or less fast and more or less far, towards the same zone of minimal egoism and minimal pleasure—to where, for the more ecstatic creature, the divine light glows with greater amplitude and greater limpidity, beyond the intermediaries which have been, *not rejected*, but *outstripped*.

Under the converging action of these three rays—purity, faith and fidelity—the world melts and folds.

Like a huge fire that is fed by what should normally extinguish it, or like a mighty torrent which is swelled by the very obstacles placed to stem it, so the tension engendered by the encounter between man and God dissolves, bears along and volatilises created things and makes them all, equally, serve the cause of union.

Joys, advances, sufferings, setbacks, mistakes, works, prayers, beauties, the powers of heaven, earth and hell—everything bows down under the touch of the heavenly waves; and everything yields up the portion of positive energy contained within its nature so as to contribute to the richness of the divine *milieu*.

Like the jet of flame that effortlessly pierces the hardest metal, so the spirit drawn to God penetrates through the world and makes its way enveloped in the luminous vapours of what it sublimates with him.

It does not destroy things, nor distort them; but it liberates things, directs them, transfigures them, animates them. It does not leave things behind but, as it rises, it leans on them for support; and carries along with it the chosen part of things.

Purity, faith and fidelity, static virtues and operative virtues, you are truly, in your serenity, nature's noblest energies—those which give even the material world its final consistency and its ultimate shape. You are the formative principles of the New Earth. Through you, threefold aspect of a same trusting adoration, "we shall overcome the world": Haec est quae vincit mundum, fides nostra.

C. *The collective progress in the divine* milieu. *The communion of saints and charity*

i. *Preliminary remarks on the "individual" value of the divine* milieu

In the foregoing pages we have been concerned in practice with the establishment and progress of the divine *milieu* in a soul envisaged as alone in the world in the presence of God. "But what about its relationship to other people?" more than one reader must have thought; "where do other people come in? What sort of Christianity is this, that thinks it can build up an edifice without regard to love of neighbour?"

Our neighbour, as will now be seen, has an essential place in the edifice whose general outline we have tried to trace. But before we could insert him within its structure, we had to deal thoroughly with the problem of the "divinisation of the world" in the particular case of an individual man; and this for two reasons.

In the first place for reasons of *method;* for, by sound scientific rules, the study of particular cases must precede an attempt at generalisation.

In the second place, for reasons of *nature;* for whatever extraordinary solidarity we have with each other in our development and in our consummation *in Christo Jesu,* each of us forms, nonetheless, a natural unit charged with his own responsibilities and his own incommunicable possibilities within that consummation. It is *we* who save ourselves or lose ourselves.

It was all the more important to stress this christian doctrine of individual salvation precisely as the perspectives developed here became more unitary and more universalist. It must never be forgotten that, as in the experimental spheres of the world, each man, though enveloped within the same universe as all other men, presents an independent centre of perspective and activity for that universe (so that there are as many partial universes as there are individuals), so in the realm of heavenly realities, however deeply impregnated we may be by the same creative and redemptive force, each one of us constitutes a particular centre of divinisation (so that there are as many partial divine *milieux* as there are christian souls).

Men, as we know, according to the dimness or excellence of their senses and intelligence, react so differently in the same circumstances and in the presence of the same opportunities of perception and action, that if *per impossibile* we could migrate from one consciousness into another we should each time change our world. In the same way, God presents and gives himself to our souls under the same temporal and spatial "species," but with very different degrees of reality and fullness, according to the faith, fidelity and purity which his influence encounters. An achievement or a disaster which involves a whole group of men has as many different facets, finalities and "souls" as there are individuals involved: blind, absurd, indifferent or material to those who do not love and do not believe, it will be luminous, providential, charged with significance and love to those who have succeeded in seeing and touching God everywhere. There are as many sur-animations by God of secondary causes as there are forms of human trust and human fidelity. Although essentially single in its influx, Providence is pluralised when in contact with us—just as a ray of sunlight takes on colour or loses itself in the depths of the body which it strikes. The universe has many different storeys and many different compartments: *in eadem domo, multae mansiones.*

That is why, in repeating over *our* lives the words the priest says over the bread and wine before the consecration, we should pray, each one of us, that the world may be transfigured for our use; *ut nobis Corpus et Sanguinis fiat D.N. Jesu Christi.*

That is the first step. Before considering others (and in order to do so) the believer must make sure of his own personal sanctification—not out of egoism, but with a firm and broad understanding that the task of each one of us is to divinise the whole world in an infinitesimal and incommunicable degree.

We have tried to show how this partial divinisation is possible. It only remains for us to integrate the elemental phenomenon and see how the total divine *milieu* is formed by the confluence of our individual divine *milieux,* and how, in order to complete them, it reacts in its turn upon the particular destinies which it clasps in its embrace. The time has come to generalise our conclusions by multiplying them to infinity by the action of charity.

ii. The intensification of the divine milieu *through charity*

In order to measure and understand the power of divinisation contained in love for one's neighbour, we must reexamine some of the themes already considered, and especially those passages in which we discussed the total unity of the eucharistic consecration.

Across the immensity of time and the disconcerting multiplicity of individuals, one single operation is taking place: the annexation to Christ of his chosen; one single thing is being made: the mystical body of Christ, starting from all the sketchy spiritual powers scattered throughout the world. *Hoc est Corpus meum.* Nobody in the world can save us, or lose us, in our despite; that is true. But it is also true that our salvation is not pursued or achieved except in *solidarity* with the justification of the whole "body of the elect." In a real sense, only one man will be saved: Christ, the head and living summary of humanity. Each one of the elect is called to see God face to face. But his act of vision will be vitally inseparable from the elevating and illuminating action of Christ. In heaven we ourselves shall contemplate God, but, as it were, through the eyes of Christ.

If this is so, then our individual mystical effort awaits an essential completion in its union with the mystical effort of all other men. The divine *milieu* which will ultimately be one in the Pleroma, must begin to become one during the earthly phase of our existence. So that although the Christian who hungers to live in God may have attained all possible purity of desire, faith in prayer, and fidelity in action, the divinisation of his universe is still open to vast possibilities. It would still remain for him to link his elemental work to that of all the labourers who surround him. The innumerable partial worlds which envelop the diverse human monads press in upon him from all around. His task is to re-kindle his own ardour by contact with the ardour of all these foci, to make his own sap communicate with that circulating in the other cells, to receive or propagate movement and life for the common benefit, and to adapt himself to the common temperature and tension.

To what power is it reserved to burst asunder the envelope in which our individual microcosms tend jealously to isolate themselves and vegetate? To what force is it given to merge and exalt our partial rays into the principal radiance of Christ?

To charity, the beginning and the end of all spiritual relationships. Christian charity, which is preached so fervently by the Gospels, is nothing else than the more or less conscious cohesion of souls engendered by their communal convergence *in Christo Jesu*. It is impossible to love Christ without loving others (in proportion as these others are moving towards Christ). And it is impossible to love others (in a spirit of broad human communion) without moving nearer to Christ. Hence automatically, by a sort of living determinism, the individual divine *milieux,* in proportion as they establish themselves, tend to fuse one with another; and in this association they find a boundless increase of their ardour. This inevitable conjunction of forces has always been manifested, in the interior lives of the saints, by an overflowing love for everything which, in creatures, carries in itself a germ of eternal life. We have already examined "the tension of communion" and its wonderful efficacy for directing man towards his human duty. It enables him to extract life even from powers which seem most heavily charged with death, and its ultimate effect is to precipitate the Christian into the love of souls.

The man with a passionate sense of the divine *milieu* cannot bear to find things about him obscure, tepid and

empty which should be full and vibrant with God. He is paralysed by the thought of the numberless spirits which are
linked to his in the unity of the same world, but are not yet
fully kindled by the flame of the divine presence. He had
thought for a time that he had only to stretch out his *own*
hand in order to touch God to the measure of his desires. He
now sees that the only human embrace capable of worthily
enfolding the divine is that of all men opening their arms to
call down and welcome the Fire. The only subject ultimately
capable of mystical transfiguration is the whole group of
mankind forming a single body and a single soul in charity.

And this coalescence of the spiritual units of creation
under the attraction of Christ is the supreme victory of faith
over the world.

I confess, my God, that I have long been, and even now
am, recalcitrant to the love of my neighbour. Just as much as I
have derived intense joy in the superhuman delight of dissolving
myself and losing myself in the souls for which I was destined by
the mysterious affinities of human love, so I have always felt an
inborn hostility to, and closed myself to, the common run of
those whom you tell me to love. I find no difficulty in integrat-
ing into my inward life everything above and beneath me (in
the same line as me, as it were) in the universe—whether mat-
ter, plants, animals; and then powers, dominions and angels:
these I can accept without difficulty and delight to feel myself
sustained within their hierarchy. But "the other man," my
God—by which I do not mean "the poor, the halt, the lame and
the sick," but "the other" quite simply as "other," the one who
seems to exist independently of me because his universe seems
closed to mine, and who seems to shatter the unity and the
silence of the world for me—would I be sincere if I did not
confess that my instinctive reaction is to rebuff him? and that

the mere thought of entering into spiritual communication with him disgusts me?

Grant, O God, that the light of your countenance may shine for me in the life of that "other." The irresistible light of your eyes shining in the depth of things has already guided me towards all the work I must accomplish, and all the difficulties I must pass through. Grant that I may see you, even and above all, in the souls of my brothers, at their most personal, and most true, and most distant.

The gift which you call on me to make to these brothers— the only gift which my heart can make—is not the overwhelming tenderness of those specially privileged affections which you have placed in our lives as the most potent created factor of our inward growth, but something less sweet, but just as real, and more strong. Between myself and men, and with the help of your eucharist, you want the foundational attraction (which is already dimly felt in all love, if it is strong) to be made manifest—that which mystically transforms the myriad of rational creatures into a sort of single monad in you, Jesus Christ. You want me to be drawn towards "the other," not by simple personal sympathy, but by what is much higher: the united affinities of a world for itself, and of that world for God.

You do not ask for the psychologically impossible—since what I am asked to cherish in the vast and unknown crowd is never anything save one and the same personal being which is yours.

*Nor do you call for any hypocritical protestations of love for my neighbour, because—since my heart cannot reach your person except at the depths of all that is most individually and concretely personal in every "other"—it is to the "other" him-*self, *and not to some vague entity around him, that my charity is addressed.*

No, you do not ask anything false or unattainable of me. You merely, through your revelation and your grace, force what is most human in me to become conscious of itself at last. Humanity was sleeping—it is still sleeping—imprisoned in the narrow joys of its little closed loves. A tremendous spiritual power is slumbering in the depths of our multitude, which will manifest itself only when we have learnt to break down the barriers *of our egoisms and, by a fundamental recasting of our outlook, raise ourselves up to the habitual and practical vision of universal realities.*

Jesus, Saviour of human activity to which you have given meaning, Saviour of human suffering to which you have given living value, be also the Saviour of human unity; compel us to discard our pettinesses, and to venture forth, resting upon you, into the uncharted ocean of charity.

iii. *The outer darkness and the lost souls*

The history of the kingdom of God is, directly, one of a reunion. The total divine *milieu* is formed by the incorporation of every elected spirit in Jesus Christ. But to say "elect" is to imply a choice, a selection. We should not be looking at the universal action of Jesus from a fully christian point of view if it were seen merely as a centre of attraction and beatification. It is precisely because he is the one who unites that he is also the one who separates and judges. The Gospel speaks of the good seed, the sheep, the right hand of the Son of Man, the wedding feast and the fire that kindles joy. But there are also the tares, the goats, the left hand of the Judge, the closed door, the outer darkness; and, at the antipodes of the fire that unites in love, there is the fire that destroys in isolation. The whole process out of which the New Earth is gradually born is an *aggregation* underlaid by a *segregation*.

In the foregoing pages (solely concerned with rising towards the divine focus and with offering ourselves more completely to its rays) our eyes have been systematically turned towards the light, though we have never ceased to feel the darkness and the void beneath us—the rarefication or absence of God over which our path has been suspended. But this nether darkness, which we sought to flee, could equally well have been a sort of abyss opening on to sheer nothingness. Imperfection, sin, evil, the flesh, appeared to us mainly as a retrograde step, a reverse aspect of things, which ceased to exist for us the further we penetrated into God.

Your revelation, O Lord, compels me to believe more. The powers of evil, in the universe, are not only an attraction, a deviation, a minus sign, an annihilating return to plurality. In the course of the spiritual evolution of the world, certain conscious elements in it, certain monads, deliberately detached themselves from the mass that is stimulated by your attraction. Evil has become incarnate in them, has been "substantialised" in them. And now I am surrounded by dark presences, by evil beings, by malign things, intermingled with your luminous presence. That separated whole constitutes a definitive loss, an immortal wastage from the genesis of the world. There is not only nether darkness; there is also outer darkness. That is what the Gospel tells us.

Of the mysteries which we have to believe, O Lord, there is none, without a doubt, which so affronts our human views as that of damnation. And the more human we become, that is to say conscious of the treasures hidden in the least of beings and of the value represented by the smallest atom in the final unity, the more lost we feel at the thought of hell. We could perhaps understand falling back into in-existence . . . but what are we to make of eternal uselessness and eternal suffering?

You have told me, O God, to believe in hell. But you have forbidden me to hold with absolute certainty that any single man has been damned. I shall therefore make no attempt to consider the damned here, nor even to discover—by whatsoever means—whether there are any. I shall accept the existence of hell on your word, as a structural element in the universe, *and I shall pray and meditate until that awe-inspiring thing appears to me as a strengthening and even blessed complement to the vision of your omnipresence which you have opened out to me.*

And in truth, Lord, there is no need for me to force either my mind or things in order to perceive a source of life even in the mystery of that second death. We do not have to peer very closely into that outer darkness to discover in it a great tension and a further deepening of your greatness.

I know that the powers of evil, considered in their deliberate and malign action, can do nothing to trouble the divine milieu *around me. As they try to penetrate into my universe, their influence (if I have enough faith) suffers the lot common to all created energy; caught up and twisted round by your irresistible energy, temptations and evils are converted into good and fan the fires of love.*

I know, too, that considered from the point of view of the void created by their defection from the mystical body, the fallen spirits cannot detract from the perfection of the Pleroma. Each soul that is lost in spite of the call of grace ought to spoil the perfection of the final and general union; but instead, O God, you offset it by one of those recastings which restore the universe at every moment to a new freshness and a new purity. The damned are not excluded from the Pleroma, but only from its luminous aspect, and from its beatification. They lose it, but they are not lost to it.

The existence of hell, then, does not destroy anything and does not spoil anything in the divine milieu *whose progress all around me I have followed with delight. I can even feel, moreover, that it effects something great and new there. It adds an accent, a gravity, a contrast, a depth which would not exist without it. The peak can only be measured from the abyss which it crowns.*

I was speaking a moment or two ago—looking at things from man's point or view—of a universe closed, from below, by nothingness, that is to say of a ladder of magnitudes that somehow stops dead at zero. But now, O God, tearing open the nether darkness of the universe, you show me that there is another hemisphere at my feet—the very real domain, descending without end, of existences which are, at least, possible.

Does the reality of this negative pole of the world not double the immensity and the urgency of the power with which you come upon me?

O Jesus, our splendidly beautiful and jealous Master, closing my eyes to what my human weakness cannot as yet understand and therefore cannot bear—that is to say, to the reality of the damned—I desire at least to make the ever present threat of damnation a part of my habitual and practical vision of the world, not in order to fear you, but in order to be more intensely yours.

Just now I besought you, Jesus, to be not only a brother for me, but a God. Now, invested as you are with the redoubtable power of selection which places you at the summit of the world as the principle of universal attraction and universal repulsion, you truly appear to me as the immense and living force which I was seeking everywhere that I might adore it: the fires of hell and the fires of heaven are not two different forces, but contrary manifestations of the same energy.

I pray, O Master, that the flames of hell may not touch me nor any of those whom I love, and even that they may never touch anyone (and I know, my God, that you will forgive this bold prayer); but that, for each and every one of us, their sombre glow may add, together with all the abysses that they reveal, to the blazing plenitude of the divine milieu.

EPILOGUE

In Expectation of the Parousia

Segregation and aggregation. Separation of the evil elements of the world, and "co-adunation" of the elemental worlds that each faithful spirit constructs around him in work and pain. Under the influence of this twofold movement, which is still almost entirely hidden, the universe is being transformed and is maturing all around us.

We are sometimes inclined to think that the same things are monotonously repeated over and over again in the history of creation. That is because the season is too long by comparison with the brevity of our individual lives, and the transformation too vast and too inward by comparison with our superficial and restricted outlook, for us to see the progress of what is tirelessly taking place in and through all matter and all spirit. Let us believe in revelation, once again our faithful support in our most human forebodings. Under the commonplace envelope of things and of all our purified and salvaged efforts, a new earth is being slowly engendered.

One day, the Gospel tells us, the tension gradually accumulating between humanity and God will touch the limits prescribed by the possibilities of the world. And then will come the end. Then the presence of Christ, which has been silently accruing in things, will suddenly be revealed—like a flash of light from pole to pole. Breaking through all the barriers within which the veil of matter and the water-tightness of souls have seemingly kept it confined, it will invade the face of the earth. And, under the finally-liberated action of the true affinities of being, the spiritual atoms of the world will be borne along by a force generated by the powers of cohesion proper to the universe itself, and will occupy, whether within Christ or without Christ (but always under the influence of Christ), the place of happiness or pain designated for them by the living structure of the Pleroma. *Sicut fulgur exit ab Oriente et paret usque ad Occidentem ... Sicut venit diluvium et tulit omnes ... Ita erit adventus Filii hominis.* Like lightning, like a conflagration, like a flood, the attraction exerted by the Son of Man will lay hold of all the whirling elements in the universe so as to reunite them or subject them to his body. *Ubicumque fuerit corpus congregabuntur et aquilae.*

Such will be the consummation of the divine *milieu.*

As the Gospel warns us, it would be vain to speculate as to the hour and the modalities of this formidable event. But we have to *expect* it.

Expectation—anxious, collective and operative expectation of an end of the world, that is to say of an issue for the world—that is perhaps the supreme christian function and the most distinctive characteristic of our religion.

Historically speaking, that expectation has never ceased to guide the progress of our faith like a torch. The Israelites were constantly expectant, and the first Christians too. Christmas, which might have been thought to turn our gaze towards the past, has only fixed it further in the future. The Messiah, who appeared for a moment in our midst, only allowed himself to be seen and touched for a moment before vanishing once again, more luminous and ineffable than ever, into the depths of the future. He came. Yet now we must expect him—no longer a small chosen group among us, but all men—once again and more than ever. The Lord Jesus will only come soon if we ardently expect him. It is an accumulation of desires that should cause the Pleroma to burst upon us.

Successors to Israel, we Christians have been charged with keeping the flame of desire ever alive in the world. Only twenty centuries have passed since the Ascension. What have we made of our expectancy?

A rather childish haste, combined with the error in perspective which led the first generation of Christians to believe in the immediate return of Christ, has unfortunately left us disillusioned and suspicious. Our faith in the kingdom of God has been disconcerted by the resistance of the world to good. A certain pessimism, perhaps encouraged by an exaggerated conception of the original fall, has led us to regard the world as decidedly and incorrigibly wicked. And so we have allowed the flame to die down in our sleeping hearts. No doubt we see with greater or less distress the approach of individual death. No doubt, again, our prayers and actions are conscientiously directed to bringing about "the coming of God's kingdom." But in fact how many of us are genuinely moved in the depths of our hearts by the wild hope that *our* earth will be recast? Who is there who sets a course in the midst of our darkness towards the first glimmer of a *real* dawn? Where is the Christian in whom the impatient longing for Christ succeeds, not in submerging (as it should) the cares of human love and human interests, but even in counter-balancing them? Where is the Catholic as passionately vowed *(by conviction* and not *by convention)* to spreading the hopes of the Incarnation as are many humanitarians to spreading the dream of the new city? We persist in saying that we keep vigil in expectation of the Master. But in reality we should have to admit, if we were sincere, *that we no longer expect anything.*

The flame must be revived at all costs. At all costs we must renew in ourselves the desire and the hope for the great

Coming. But where are we to look for the source of this rejuvenation? We shall clearly find it, first and foremost, in an increase of the attraction exercised directly by Christ upon his members. And then *in an increase of the interest,* discovered by our thought, in the preparation and consummation of the Parousia. And from where is this interest itself to spring? From the perception of *a more intimate connection* between the victory of Christ and the outcome of the work which our human effort here below is seeking to construct.

We are constantly forgetting that the supernatural is a ferment, a soul, and not a complete and finished organism. Its role is to transform "nature"; but it cannot do so apart from the matter which nature provides it with. If the Jewish people have remained turned towards the Messiah for three thousand years, it is because he appeared to them to enshrine the glory of their people. If the disciples of St. Paul lived in perpetual expectation of the great day, that was because it was to the Son of Man that they looked for a personal and tangible solution to the problems and the injustices of life. The expectation of heaven cannot remain alive unless it is incarnate. What body shall we give to ours today?

That of a huge and *totally human* hope. Let us look at the earth around us. What is happening under our eyes within the mass of peoples? What is the cause of this disorder in society, this uneasy agitation, these swelling waves, these whirling and mingling currents and these turbulent and formidable new impulses? Mankind is visibly passing through a crisis of growth. Mankind is becoming dimly aware of its shortcoming and its capacities. And as we said on the first page, it sees the universe growing luminous like the horizon just before sunrise. It has a sense of premonition and of expectation.

Subject, like everyone else, to that attraction, the Christian, we said, sometimes wonders, and is uneasy. May he not be bestowing his adoration on an idol?

Our study, now completed, of the divine *milieu* suggests an answer to this fear.

Those of us who are disciples of Christ must not hesitate to harness this force, which needs us, and which we need. On the contrary, under pain of allowing it to be lost and of perishing ourselves, we should share those aspirations, in essence religious, which make the men of today feel so strongly the immensity of the world, the greatness of the mind, and the sacred value of every new truth. It is in this way that our christian generation will learn again to expect.

We have gone deeply into these new perspectives: the progress of the universe, and in particular of the human universe, does not take place in competition with God, nor does it squander energies that we rightly owe to him. The greater man becomes, the more humanity becomes united, with consciousness of, and master of, its potentialities, the more beautiful creation will be, the more perfect adoration will become, and the more Christ will find, for mystical extensions, a body worthy of resurrection. The world can no more have two summits than a circumference can have two centres. The star for which the world is waiting, without yet being able to give it a name, or rightly appreciate its true transcendence, or even recognise the most spiritual and divine of its rays, is, necessarily, Christ himself, in whom we hope. To desire the Parousia, all we have to do is to let the very heart of the earth, as we christianise it, beat within us.

Men of little faith, why then do you fear or repudiate the progress of the world? Why foolishly multiply your warnings and your prohibitions? "Don't venture . . . Don't try . . . everything is known: the earth is empty and old: there is nothing more to be discovered."

We must try everything for Christ; we must hope everything for Christ. *Nihil intentatum*. That, on the contrary, is the true christian attitude. To divinise does not mean to destroy, but to sur-create. We shall never know all that the Incarnation still expects of the world's potentialities. We shall never put enough hope in the growing unity of mankind.

Jerusalem, lift up your head. Look at the immense crowds of those who build and those who seek. All over the world, men are toiling—in laboratories, in studios, in deserts, in factories, in the vast social crucible. The ferment that is taking place by their instrumentality in art and science and thought is happening for your sake. Open, then, your arms and your heart, like Christ your Lord, and welcome the waters, the flood and the sap of humanity. Accept it, this sap—for, without its baptism, you will wither, without desire, like a flower out of water; and tend it, since, without your sun, it will disperse itself wildly in sterile shoots.

The temptations of too large a world, the seductions of too beautiful a world—where are these now?

They do not exist.

Now the earth can certainly clasp me in her giant arms. She can swell me with her life, or take me back into her dust. She can deck herself out for me with every charm, with every horror, with every mystery. She can intoxicate me with her perfume of tangibility and unity. She can cast me to my knees in expectation of what is maturing in her breast. . . .

But her enchantments can no longer do me harm, since she has become for me, over and above herself, the body of him who is and of him who is coming.

The divine milieu.

Tientsin,
November 1926–March 1927

FRENCH EDITOR'S NOTE

In March 1955, the last month of his life, Père Teilhard de Chardin's thoughts went back to *Le Milieu Divin,* and he wrote at the beginning of a final profession of faith:

> It is a long time now since, in *La Messe sur le Monde* and *Le Milieu Divin,* I tried to put into words the admiration and wonder I felt as I confronted perspectives as yet hardly formulated within me.
>
> Today, after forty years of constant reflection, it is still exactly the same fundamental vision which I feel the need to set forth and to share, in its mature form, for the last time. With less exuberance and freshness of expression, perhaps, than at my first encounter with it, but still with the same wonder and the same passion.

No work of this great believer can be understood except in relation to this "fundamental vision" of *Le Milieu Divin*—the vision (always implicit, even when not stated) of Christ as *All-in-everything;* of the universe moved and com-penetrated by God in the totality of its evolution.

Index

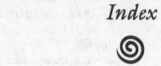

Perennial

Books by Pierre Teilhard de Chardin:

THE PHENOMENON OF MAN
ISBN 0-06-090495-X (paperback)

Pierre Teilhard de Chardin applied his whole life, his tremendous intellect, and his great spiritual faith to building a philosophy that would reconcile Christian theology with the scientific theory of evolution—to relate the facts of religious experience to those of natural science. *The Phenomenon of Man* contains the quintessence of his thought.

"A most extraordinary book, of far-reaching significance for the understanding of man's place in the universe."—Abraham J. Heschel

THE DIVINE MILIEU
ISBN 0-06-093725-4 (paperback)

The Divine Milieu is an essential companion to Teilhard de Chardin's *The Phenomenon of Man*. In it he expands on the spiritual message so basic to his thought and shows how man's spiritual life can become a participation in the destiny of the universe.

"Extraordinary." —Karl Stern

Available wherever books are sold, or call 1-800-331-3761 to order.